H. W. Long

Sane Sex Life and Sane Sex Living

H. W. Long

Sane Sex Life and Sane Sex Living

Reproduction of the original.

1st Edition 2022 | ISBN: 978-3-36826-476-5

Verlag (Publisher): Outlook Verlag GmbH, Zeilweg 44, 60439 Frankfurt, Deutschland
Vertretungsberechtigt (Authorized to represent): E. Roepke, Zeilweg 44, 60439 Frankfurt, Deutschland
Druck (Print): Books on Demand GmbH, In de Tarpen 42, 22848 Norderstedt, Deutschland

NOTE TO THE READER

IN ORDER TO GAIN A CORRECT IMPRESSION OF THE BOOK, IT IS ESSENTIAL THAT IT BE READ FROM THE BEGINNING TO THE END WITHOUT ANY SKIPPING WHATSOEVER. ONCE READ, IT CAN BE RE-READ, HERE AND THERE, AS THE READER MAY DESIRE. BUT FOR A FIRST READING, IT IS THE EARNEST WISH OF THE AUTHOR THAT EVERY WORD BE READ, FOR IN NO OTHER WAY CAN THE PURPOSE OF THE BOOK BE REALIZED.

INTRODUCTION

As we have moved down the ages, now and then, from the religious teacher, the statesman, the inventor, the social worker, or from the doctor, surgeon, or sexologist, there has been a "*vox clamantis in deserto.*" Usually these voices have fallen on unheeding ears; but again and again some delver in books, some student of men, some inspired, self-effacing, or altruistic one has taken up the cry; and at last unthinking, unheeding, superficial, self-satisfied humanity has turned to listen.

Aristotle by the sure inductive method learned and taught much, concerning the sex relations of men and women, that it would profit us today to heed. Balzac, Luther, Michelet, Spencer, and later, at our very doors, Krafft-Ebbing, Forel, Bloch, Ellis, Freud, Hall, and scores of others have added their voices. All these have seen whither we were drifting, and have made vigorous protests according to their lights. Many of these protests should have been heard, but were not, and only now are just beginning to be heeded. Such pioneers in the field of proper, healthful, ethical, religious, sane daily sex living, have been Sturgis and Malchow, who talked earnestly to an unheeding profession of these things, and now, I have the honor to write an introductory word to a book in this field, that is sane, wise, practical, entirely truthful, and unspeakably necessary.

I can endorse the teachings in Dr. Long's book more fully because I have, for nearly a quarter of a century, been holding similar views, and dispensing similar, though perhaps less explicit, information. I know from long observation that the teaching is wholesome and necessary, and that the results are universally uplifting. Such teachings improve health, prolong life, and promote virtue, adding to the happiness and lessening the burdens of men, on the one hand; on the other, reducing their crimes and vices. A book like this would have proved invaluable to me on my entrance to the married state; but had I had it, I might not have been forced to acquire the knowledge which enables me now to state with all solemnity, that I personally know hundreds of couples whose lives were wrecked for lack of such knowledge, and that I more intimately know hundreds of others to whom verbal teaching along the lines he has laid down, has brought happiness, health and goodness.

Dr. Long advances no theories; neither do I. He has found by studying himself and other people, a sane and salutary way of sex living, and fearlessly has prescribed this to a limited circle for a long time. I congratulate him for his perspicacity, temerity, and wisdom. He offers no apology, and there is no occasion for any. He says, "All has been set down in love, by a lover, for the sake of lovers yet to be, in the hope of helping them on toward a divine consummation." That is, he has developed these ideas at home, and then spread them abroad, or, he has found them abroad and brought them home; and they worked.

I also speak somewhat *ex experientia* and have some intimate personal knowledge of many of these things. Therefore, I advocate his doctrine, the more readily, and maintain that humanity needs these ideas as much today as when M. Jules Lemaitre wrote his late introduction to Michelet's *L'Amour*. He said: "*Il ne parait pas, apres quarante ans passes, que les choses aillent mieux, ni que le livre de Michelet ait rien perdu de son a-propos.*" Twenty years more have elapsed and things have not yet become much better. Frank sex talks like

Dr. Long's teaching are as a-propos today as was Michelet's book when it was written, or when, after forty years had passed M. Lemaitre wrote his introduction.

Idealism is right, and we all approve it; so much so, that many of us cannot see that ultra-idealism, extremism in right, (it is foolish to attempt to attain anything better than the best) may be wrong. Undoubtedly, entire devotion to the material and physical, is also wrong; but we never must lose sight of the palpable fact that, unless we have a proper, stable, natural, well-regulated physical or material foundation, we must fall short of all ideals. Proper physical adjustments enable the realization of realizable ideals. Unrealizable ideals are chimeras pursued into futurity, while a world that should be human and happy waits in vice and misery. I gather that Dr. Long believes that reducing this vice and misery, and increasing human happiness and improving health are suitable works with which to companion a faith in the Arbiter of our destinies.

If thus he develops his idea of the integrity of the universe, I agree with him fully. His book, since it delineates the numerous details of a normal sex life, can be sold, thanks to our prudish public, only to the profession. I believe it should go to the larger public as it has gone formerly to his smaller community.

In spite of imperfect ideals the Orient has endured, while we of the Occident are fast becoming decadent. We, by learning something of the art of love, and of the natural life of married people, from the Hindoos, may perpetuate our civilization. They, by adopting the best of our transcendentalism, may reach higher development than we yet have attained.

The time has come for a book like this to command the attention of medical men, since now an awakened public demands from them, as the conservers of life and the directors of physiological living, explicit directions in everything pertaining to the physician's calling, not omitting the intimate, intricate, long taboo and disdained details of sex life and procreation.

W.F. ROBIE, M.D.

CONTENTS

FOREWORD

To Members of the Medical Profession into Whose Hands This Book May Come:

The following pages are more in the nature of a manuscript, or heart-to-heart talk between those who have mutual confidence in each other, than of a technical, or strictly scientific treatise of the subject in hand; and I cannot do better, for all parties concerned, than to explain, just here in the beginning, how this came about, and why I have concluded to leave the copy practically as it was originally written.

In common with nearly all members of our profession who are engaged in the general practice of medicine, I have had numbers of married men and women, husbands and wives, patients and otherwise, who have come to me for counsel and advice regarding matters which pertain to their sex-life, as that problem presented itself to them personally. As we all know, many of the most serious and complicated cases we have to deal with have their origins in these delicate relations which so often exist among wedded people, of all classes and varieties.

For a number of years I did what I could for these patrons of mine, by way of confidential talks and the like, my experience in this regard probably being about on a par with that of my medical brethren who are engaged in the same kind of work. It is needless to say that I found, as you have doubtless found under the same conditions, many obstacles to prevent satisfactory results, by this method of procedure. My patients were often so reticent, or timid and shamefaced, that it was frequently difficult to get at the real facts in their cases, and, as we all know, many of these would, for these and other reasons, conceal more than they revealed, thereby keeping out of evidence the most vital and significant items in their individual cases. All these things, of course, tended to make bad matters worse, or resulted in nothing that was really worth while.

After some years of this sort of experience, and meditating much on the situation, I came to the conclusion that a very large percentage of all this trouble which I and my patrons had to go up against, was almost entirely the result of ignorance on the part of those who came to consult me; and because knowledge is always the antidote for not knowing, I came to the conclusion that, if it were possible to "put these people wise" where they were now so uninformed, I might at once save them from a deal of harm and myself from much trouble and annoyance.

Further than this, I remembered once hearing a wise man say that often "what cannot be said may be sung"; and I realized that it is equally true that much which would be awkward, or embarrassing, if said to a person, face to face, might be got to them in writing with impunity. This I found to be especially true of my women patients, some of whom might become suspicious of a wrong intent from the things said in a private conversation, when they would have no such fears or doubts if they read the same words from a printed page. It was these considerations which first suggested to me the writing of the following pages.

Still other reasons why I did as I did were as follows: You see, at once, if you stop to think about it, that the writing out of the knowledge I proposed to impart was really a matter of necessity for me, because of the *saving of time* which would thereby be secured. To get any results that would be worth while in these matters, I would be required to tell about ever so many things concerning which they were totally ignorant; and to tell about ever so many things, by word of mouth, to each individual patient, *takes time* — ever so much time, if the work is well done, and it had better not be done at all if it is not well done. So I really was forced to write out what I wanted to teach these patients of mine.

And let me say further that I was compelled to write these things out for my people as I have written them, because, in all the range of literature on this vital subject, I knew of nothing which would tell them just what it seemed to me they ought to be told, and what they ought to know.

And so it was that I wrote the manuscript which is now printed in the following pages. I did not write it at first just as it now stands, because experience showed me, from time to time, where my first efforts could be modified and improved. So what is here presented is the result of many practical demonstrations of the real working value of what the manuscript contains.

My method of using the copy has been something as follows: As I have already suggested, what I have written has been prepared for the sole and express purpose of helping husbands and wives to live sane and wholesome sex-lives — to give them the requisite knowledge for so doing; knowledge of themselves and of each other as sexual beings; the correct ideas regarding such right manner of living; to disabuse their minds of wrong sex-teaching, or no teaching at all, of ignorance, or prudery, or carelessness, or lust — in a word, to get to them the things that all sane married people ought to know, and to help them to practice these things, to the best of their several abilities.

(Perhaps I ought to say that there is not a line of what I have written that deals with the subject of venereal diseases, any of them. This field is already so well covered by a literature especially devoted to this subject that it needs no word of mine to make it as satisfactory as possible, as far as discoveries regarding the same have progressed. My attempt is toward making marriage more of a success than it now is, under existing conditions; and we all know that there is a limitless field for exploration and exploitation right there.)

Speaking somewhat generally, I have found what I have written to be of special value to two classes of my patrons: First, to the "newly-weds"; and, second, to those who have been married for a longer or shorter period, and who "have not got on well together." A word or two regarding each of these:

It is a wise old saying that "an ounce of prevention is worth a pound of cure," and in no other experience of life is this so true as in the ills to which married people are peculiarly subject. Many a newly wedded couple have wrecked the possibilities of happiness of a life time on their "honeymoon trip"; and it is a matter of common knowledge to the members of our profession that the great majority of brides are practically raped on their entrance into the married relation. Further than this, we all know that these things are as they are chiefly because of the ignorance of the parties concerned, rather than because they deliberately meant to do wrong. They were left to travel, alone and unguided, over what was to them an unknown way, one that was beset with pitfalls and precipices, and where dangers lurked in every forward step they took. It is to these that I have found what I have written to be a great help at the time of their utmost need; and the thanks I have received from such parties have been beyond the power of words to express.

As to just when it is best to put this information into the hands of young married people, my experience has varied with the personality of the parties concerned. In some cases I have put the copy into their hands some time before their marriage; in others, not till some time thereafter; but, as a rule, I have got the best results by putting the manuscript into their hands just at the time of their marriage, and in most of these cases the greatest success has come from their reading it together during their honeymoon. However, this is a matter on which I do not care to advise, and regarding which each practitioner must act to the best of his own judgment.

Once more: Because it is not safe to assume that young married people are already possessed of the *details* of the essential knowledge which they ought to possess, and because such *details* are the *very heart* of the whole matter, I have made these details as simple and explicit as possible, more so than might seem necessary to the professional reader. But my experience has proven that I was wise in this regard, as these very details have saved the day in more than one case, as the parties who have reported to me, after having read what I have written, have frequently testified. Sometimes a bride and groom would keep the copy for a few days only, giving it but a single reading; but, as a rule, they have been anxious to retain it for some time, and to read it again and again, especially some parts of it, till they were well posted on all that it contains. I found, too, that those who had received help from the reading of the manuscript were glad to tell others of their friends of the benefits they had received, and that thus there was a constantly widening circle.

Of course, not all young married people are capable of reading this book with profit to themselves or anyone else; but many of them are, and these ought to have the privilege of doing so. Your own good sense and experience will determine who these latter are, and these you can favor as they deserve. It is because of this situation that this book can only be used professionally that it needs the guiding hand of an expert physician to insure its reaching only those who can be benefited by its reading.

As to the other class of readers, those who have not got on well in the marriage relation (and we all know that the name of these is legion) my experience in getting to them what I have written has been quite varied; but, on the whole, the results have been good — many times they have been most excellent. Of course, it is harder to correct errors than to prevent them; but as most of the errors I have had to deal with among this class of patients have been made through ignorance rather than otherwise, I have found that the establishment of knowledge in the premises has generally brought relief where before was only suffering and woe.

Another way in which I have found the copy to be of the greatest value with these cases of unsatisfactory marital relations is the fact that, often, by the parties *reading the copy together* they have come to a mutual understanding by so doing, and have established a *modus vivendi* which could not have been attained in any other way. When such parties see their doctor singly, either of them, a prejudiced view is very apt to result, and they would seldom, if ever, come together to consult a physician regarding their troubles. But the *reading of the book together* makes a condition of affairs which is very apt to work out for the best interests of all parties concerned. Certainly, this is true, that in no case has the reading of the book made bad matters worse, and in many cases, (indeed in nearly all of them) it has been of untold value and benefit to the readers.

And because these things are so, because what I have written has proved its worth in so many cases, I have finally concluded to give the copy a larger field in which it may be used by other members of the profession besides myself. I confide it to my fellow-members in the profession feeling sure that they will use it among their patients with wisdom and discretion; and my hope is that their so doing may yield for them and theirs the most excellent results which have come to me and mine, on these lines, in the years that have gone by.

Perhaps I ought to say that the somewhat unique typography of the book, the large percentage of italics, and not a few capitalized words that appear in the pages, comes from a duplication of the copy I have used with my patients. I wrote the original copy in this way for the sake of giving special emphasis to special points for my readers, and the results attained I believe were very largely due to the typographically emphatic form of the book. Appearing in type in this way, it gives a sort of personal touch to what is thus presented to the eye of the reader, and the tendency of this is to establish a heart-to-heart relation between the author and the reader which could not be attained in any other way.

All through the copy I have avoided the use of technical words, never using such a term without explaining its meaning in plain English in the words that immediately follow it. I found this an absolute necessity in writing so that the lay reader could understand, in saying things that would produce results.

I might say, also, that the "Introduction" to the real subject matter of the book, I found necessary to write as it is largely to get my readers into a proper *mental attitude* for a reasonable recognition and understanding of what follows it. There are so many wrong teachings and biased ideas in the premises that these had to be counteracted or removed, to a degree, at least, before the rest of the copy could be rightly read. My experience is, that the preface, as it stands, has been the means of putting the readers of the book into a right mental attitude for its successful study and consideration. For the good of the cause it is written to serve, and for help to those who need help in the most sacred and significant affairs of their lives, may the book go on its way, if not rejoicing in itself, yet causing rejoicing in the lives and hearts of all who read what its pages contain.

H.W.L.

I
AN EXPLANATORY INTRODUCTION

A pious Christian once said to me: "I find it hard to reconcile sex with the purity of Providence." He never could understand why God arranged for sex anyway. Why something else might not have been done. Why children might not have come in some other fashion.

Look at the harm sex has involved. Most all the deviltry of history that was not done for money was done for sex. And even the deviltry that was done and is done for money had, and has sex back of it. Take sex out of man and you have something worth while. God must have been short of expedients when God, in sex, conceived sex. It certainly looks as if the Divine fell down this time. As if infinity was at the end of its tether. As if the adept creator for once was caught napping, or for once botched a job.

So we had my pious friend. And we had medievalism. And we had the ascetics. And heaven knows what else. Too much sex some places. Too little sex other places. Some people swearing on and some swearing off. The prostitute giving away that which was meant to be kept. The virgin keeping that which was meant to be given away. A force contending with a force. Drawing in opposite directions when they should be pulling together. Through it all, motherhood misunderstood. And fatherhood misunderstood. The body cheapened to the soul. And the soul cheapened to the body. Every child being a slap in the face of virtue.

Have you ever tried to see what this came from and goes to? This philosophy of vulgar denial? This philosophy of wallowing surrender?

The Christian stream has been polluted. It has gone dirty in the age of hush. We are supposed to keep our mouths shut. We are not to give sex away. We breed youngsters in fatal ignorance. They are always asking questions. But we don't answer their questions. The church don't answer them. Nor the state. Nor the schools. Not even mothers and fathers. Nobody who could answer answers them. But they don't go unanswered. They get answered. And they get answered wrong instead of right. They get answered, smutched instead of washed. They get answered blasphemously instead of reverently. They get answered so that the body is suspected instead of being trusted.

A boy who knows nothing asks a boy who knows nothing. A girl who knows nothing asks a girl who knows nothing. From nothing nothing comes. Men who have been such boys know nothing. Women who have been such girls know nothing. From nothing nothing comes. They have become familiar with sex circumstances. They are parents. They have done the best they knew how. But they never learned sex. They never realized its fundamentals. They never went back to, or forward to it. They were lost in a wilderness. They existed without living. They took sex as they took whiskey. They breathed an atmosphere of hush. They had got past the ascetics. But they had not got to be men and women. They didn't refuse sex. But though embracing its privileges, they still seemed to regard it as something not to be gloried in. The least said about it the soonest mended. Mothers and fathers would say to children: "You'll know about it soon enough." Teachers would say: "Ask your questions at home." Home would say: "What ever started you thinking about such things?"

The child goes about wondering. What's the matter with sex that everybody's afraid to talk about it? What's the matter with my body that I dare not mention it? My body seems very beautiful to me. I like to look at it. I like to feel it. I like to smell it. But I'm always hurried into my clothes. My body is so mysteriously precious I must take care of it. But how am I to take care of it if I don't get acquainted with it?

I find that having a body has something to do with being a father and a mother. I want to be a father. I want to be a mother. But how can I be a father or mother if some one who knows doesn't tell me what precedes fatherhood and motherhood? I should prepare for it. How can I if all the books are closed? How can I if I am blanked every time I express my curiosity? Is there no one anywhere who'll be honest with me?

If I look at sex right out of my own soul, it seems like something which God didn't fail with, but succeeded with. Like something not polluted, but purified. Like something having everything, instead of only an occasional thing, to do with life. But the world shakes its head. The world is nasty. But it puts on airs. The world has eaten. But the world says it's best to starve. Folks will say they've got to be parents. But they say they will regret it. They say sex is here. They say we're up against its mandates or its passions. But let's be as decent as we can with the indecent. Let's not linger on its margins. Let's not overstay our dissipation. Sex is like eating. Who would eat if he didn't have to? To say you enjoy a meal is carnal. To say that you derive some sense of ecstasy from paternal and maternal desires is a confession of depravity. Sex at the best is a sin.

Sex at the best is like stepping down. That sex might be an ascent. That sex might be the only means of growth and expansion. You never suppose that! You only assume perdition. You are afraid to assume heaven. I may take pride in that which I may abstract from my anatomy. I must not allude to my body as frankly as to my soul. I must withdraw my body from the public eye. From discussion. From its instinctive avowals. Our bodies must be coffined. Treated as dead before they are born. Regarded as conveniences. Not as essential entities. The body is only for a little while. The soul is forever. But why is that little while not as holy as forever? They don't say. They cavalierly settle the case of the body against itself.

So it goes. Endless vivid portrayals could be made of the anomalous situation. The more you look at the mess we've got sex into the worse it seems. *Someone's got to peach.* Someone's got to tell the truth. In a world of liars who are hushers? In a world of hushers who are liars? *Someone's got to tell the truth.* Someone's got to give sex its due. *You can't give spirit its due until you give sex its due.* You can't accept one and cast aside one. They go together. They are inseparable.

You refer to body and soul as if you knew just where one stops and the other commences. Maybe neither stops and neither commences. Maybe they are not two things but two names. Maybe when you put a body into a grave you put a soul there too. And maybe you put neither there. It's not so easy to say.

I can't see anything in the things you call spiritual more marvelous than what you call the physical birth of a baby from a mother. Maybe you know all about it. I don't. I know nothing about it. To me it's mysterious. To me it's the supreme demonstration of the spiritual.

How that a baby comes from a man and a woman. I want that kept clean. It starts clean. Why do we corrupt it? You who disparage it corrupt it. You ascetics anywhere. You libidinous roues anywhere. You corrupt it. By your excesses. You who never say yes. You who never say no. You corrupt it.

You parents. You professors. You prudes. This is addressed to you. What have you got to say about it? You have tremblingly closed the question. I would coolly open it. You have rebuked God by silence. I would praise God by speech.

II
The Argument And The Information

No apology is offered for what is said in the following pages, but a brief explanation is virtually necessary to make clear, from the outset, the reasons why it has been written.

It is one of the chief characteristics of the human race that the knowledge acquired by one generation can be passed on to the generations that follow; and that, in this way, progress in the betterment of life's results and the adaptation of means to ends can make a steady and reliable advance.

Such a method of evolution and growth is not possible in the vegetable or animal kingdom, where *instinct* is the only means for the transmission of acquired knowledge. It is this feature that differentiates man from all other created beings.

But here is a curious fact: In one realm of human experiences, in all Christian civilized countries, it has been considered wrong, even in some cases being counted a criminal offense, punishable by fine and imprisonment, for anyone to make any record of, or transmit to anyone else, any knowledge that may have been acquired regarding sex relations in the human family.

To be sure, there has been preserved, from time to time, a body of *professional* knowledge of this sort, made and prepared by physicians, but *confined strictly to that class of people.* No attempt has been made to disseminate such knowledge among those who most need it — the common people. On the contrary, every possible effort is put forth to keep such knowledge from them. This is wholly at variance with the practice regarding all other forms of human knowledge, which is to spread, as widely as possible, all known data that have so far been obtained.

There is not space, in this small volume, for pointing out the reasons for this anomalous condition of affairs, but the chief cause of its status, past and present, is grounded on two sources: The first of these is a brutal selfishness which has come over to modern times from a savage past; the second is a sort of pious prudery.

The result of these causes has been to make the whole subject of sex in the human family, with its functions and mission in human affairs, together with its proper training, discipline and exercise — to make all these things *tabu*, something to be ashamed of and ignored as much as possible, and all the knowledge regarding them that one generation has been permitted to transmit to those who come after, may be summed up in these words, namely "*Thou shalt not.*"

Now it goes without saying that, in the very nature of things, *all* this is just as bad as it can possibly be. For, of all phenomena with which the human race has to do, that of the highest importance, so far as the well-being of the race is concerned, is that which has to do with sex in men and women. A large percentage of all the physical ailments in mankind and womenkind arise from errors in sexual life, and these are but trifles compared with the mental and spiritual disasters which come upon humanity from the same source. It is probably true that more than one-half of all the crimes that are committed in the civilized world are more or less directly connected with sex affairs, and there is no so common a cause for insanity as sex aberrations.

And nearly all these ills, crimes and misfortunes arise because of *ignorance* in the matter of sex in which the rank and file of the race are forced to live. Few of these ever acquire any positive and definite knowledge in the premises, and if they do learn anything for sure, *they keep it to themselves,* inspired to do so by a false belief regarding the rightful transmission of such knowledge; or, by a false modesty, or prudery, they are kept from telling to anyone else what they have discovered or found to be the truth in these matters. And so the people stumble along in ignorance of these vital affairs in life, generation after generation, repeating the errors of their predecessors, and no positive progress being made as the years go by. Because of this state of affairs millions of human beings die every generation, and other millions suffer the tortures of the damned while they live, when they should enjoy the delights of the elect, and would do so if they only knew the actual facts in the case, and would act in accordance with the knowledge that ought to be made theirs.

But there are not wanting signs of the times that there will slowly come a change in these conditions. The fact is that the intelligent world is beginning to emerge from a condition of conformity to the say-so of some one supposed to speak with authority, and to come into a realm of obedience only to a law that has a scientific basis of actual knowledge for its foundation.

For untold ages the sex relations of the human family have been directed and determined by the clergy and by *their* teachings and pronunciamentos regarding what was fit and right. There is no need of saying hard things about such a fact; nevertheless, it is true that, for the most part, all the dicta of these men have originated amongst those who knew nothing of the *scientific* conditions regarding the subject on which they issue their mandates. So did the blind lead the blind, and the ditches of the past years are filled to overflowing with the dead bodies and souls of men and women, who, for this cause, have fallen therein.

This must not always be! It is neither wise nor right that the essential matters of human life should always remain a stumbling block and a rock of offense for the children of men. We are coming to see that sex is no more unclean and to be denied a scientific knowledge of, than any other part of the human body — the eye, the ear or whatsoever. Furthermore, the rank and file are beginning to clamor for a knowledge of these matters for themselves. This is shown by the frequency of articles that deal with sex in many of the best newspapers and magazines in the civilized world, and by similar discussions in the literature, the works and scientific books that now go into the hands of the common people. It also shows in the attempts that are occasionally being made to introduce the subject of sexual hygiene into our public schools and other educational institutions. "The world do move!"

It is for these reasons — because it is right to transfer to you and to those who come after, the sex knowledge that has been acquired by the author, by reading scientific and professional literature upon the subject, by conference with men and women who know, and by personal and professional experience, that what follows is written.

III
THE CORRECT MENTAL ATTITUDE

So much by way of general remarks regarding the subject in hand. It is more the especial purpose of what follows, however, to treat of the matter of marriage in particular, *to say something definite to young husbands and wives that shall be of real benefit to them,* not only by way of starting them out right in the new and untried way upon which they have entered, but to help them to make that way a realm of perpetual and ever increasing joy to both parties concerned, throughout its entire course, their whole lives long.

Be it said, then, first, that it is the duty of every bride and groom, before they engage in sexual commerce with each other, to acquaint themselves thoroughly with the anatomy and physiology of the sex organs of human beings, both male and female, and to make the acquirement of such knowledge as dispassionate and matter-of-fact an affair as though they were studying the nature, construction and functions of the stomach, or the digestive processes entire, or the nature and use of any of the other bodily organs. "Clear and clean am I within and without; clear and clean is every scrap and part of me, and no part shall be held more sacred or preferred above another. For divine am I, and all I am, or contain."

Now the normal young man or woman would do just this, would pursue a study of sex in this way, were it not for the fact that they have been taught, time out of mind, that to do this is immodest, not to say indecent or positively wicked. They have longed to be possessed of such knowledge, all their lives; in most cases more than any other form of wisdom that it was possible for them to make their own. But its acquirement has been placed beyond their possible reach, and it is only by the most clandestine and often nasty means that they have attained what little they know. But the quotation made in the last paragraph, sounds the key note of what is *right* in this matter, and the first effort made by the reader of these pages should be to establish in himself or herself the *condition of mind which these lines embody.*

And it had better be said, right here, that for most young people this will be found to be no *easy* thing to do. Nor should the reader feel ashamed or chagrined, or at odds with himself or herself if he or she finds such condition of affairs existing in his or her case. For it is nothing for which they are to blame. It is a misfortune and not a fault. It is only the result of inherited and inculcated (the word inculcated means *kicked in*) ideas to which all "well bred" youths have been subjected for centuries; the idea being that the closer they were kept in the realm of innocence, which is only another name for ignorance, the better "bred" they are. And to pry one's self loose, to break or tear one's self away from such a mental view and condition as heredity and such years of rigorous restraint have developed, is no small task. Indeed, it often takes months, and sometimes years, wholly to rid one's self of these deep seated and powerful, wrong views and prejudices.

Remember this: that *to the pure all things are pure.* But do not make the mistake of thinking that this much abused sentence means that purity means *emptiness*! It does no such thing. On the contrary, it means *fullness*, to *perfection*. It means that one should be possessed of the right kind of stuff, and that the stuff should be of supreme quality. So, in studying to obtain a knowledge of sex organs and sex functions, in the human family, the reader should not try to divest himself or herself of all sex-passion and desire; but, on the contrary, to make these of a sort of which he or she can be *proud*, rather than *ashamed* of, rejoice in, rather than suffer from.

So, then, let the reader of these lines, first, get a correct *mental attitude* toward what is about to be said. Banish all prurient curiosity, put aside all thought of shame or shock, (these two will be hardest for young women to overcome, because of their training in false modesty and prudishness) and endeavor to approach the subject in a reverent, open-eyed, conscientious spirit, as one who wishes, above everything else, to know the honest truth in these most essential matters that pertain to human life. Get into this frame of mind, and *keep in it*, and what is here written will be read with both pleasure and profit.

Once more, for we must make haste slowly in these delicate affairs, if the reader should find himself or herself unduly excited, or perhaps shocked, while reading some parts of what is here written, so that the heart beats too fast, or the hand trembles, it may be well to suspend the reading for a time, divert the mind into other channels for a while, and resume the reading after one has regained poise and mastery of one's self. That is, "*keep your head*" while you read these lessons, and you will be all right.

IV
The Sex Organs

And now, having given these cautionary directions, the way is clear for the making of definite statements and the giving of positive instruction.

Here, then, is a brief description of the sex organs in man and woman. At first, only the names of the parts will be given, with such slight comments and explanations as are necessary for making this part of the subject clear. A detailed setting forth of the functions and proper exercise of these organs will be given later.

The sex organs in a male human being consists, broadly speaking, of the penis and the testicles. All these are located at the base of the abdomen, between the thighs and on the forward part of the body. The penis is a fleshy, muscular organ, filled with most sensitive nerves, and blood vessels that are capable of extension to a much greater degree than any of their similars in other parts of the body. In a quiescent, or unexcited condition, in the average man, this organ is from three to four inches long and about an inch or more in diameter. It hangs limp and pendent in this state, retired and in evidence not at all. In its excited, or tumescent condition (the word tumescent means swelled, and is the technical word for describing the erect condition of the penis) it becomes enlarged and rigid, its size in this state being, on an average, six or seven inches long, and from an inch-and-a-half to two inches in diameter. It is almost perfectly cylindrical, slightly thicker at the base than at its forward part.

The testicles are two kidney shaped glands, not far from the size of a large hickory nut, and are contained in a sort of sack, or pocket, called the scrotum, which is made for their comfortable and safe carrying. The scrotum hangs directly between the thighs, at the base of the penis, and in it are the testicles, suspended by vital cords that are suspended from the body above. The left testicle hangs a little higher in the sack than the right, so that, in case the thighs are crowded together, one testicle will slip over the other, and so the danger of crushing them will be avoided. This is one of the many ways which the Maker of the human body has devised to insure the proper preservation of the vital organs from harm, a fact which should inspire all human beings with profound reverence for this most wonderful of all life forms, the beautiful human body, the "temple of the Holy Spirit."

The part of the body upon which the sex organs, male and female, are located is known as the pubic region. It is covered with hair, which, in both sexes, extends well up the lower belly. This is known as pubic hair, and in general corresponds in quality and quantity to the hair upon the individual head, being coarse or fine, soft or bristly, to match, the head covering, in each case. This hair is usually more or less curly, and forms a covering an inch or more in depth over the whole pubic region, extending back between the thighs slightly beyond the rectum. In occasional cases this hair is straight and silky, and sometimes grows to great length, instances being known, in some women, where it has extended to the knees. A well-grown and abundant supply of fine pubic hair is a possession highly prized by women, of which they are justly proud, though few of them would acknowledge the fact, even to themselves. None the less it is a fact.

The female sex organs, speaking generally also, are as follows: The vulva, or outward portion of the parts; the vaginal passage; the uterus, or womb, and the ovaries. All but the first named lie within the body of the woman. The vulva is made up of several parts which will be named and described later. The vaginal passage is a tube, or canal leading from the vulva to the womb. In length and diameter it corresponds almost exactly with that of the penis, being six or seven inches in depth, and capable of a lateral extension which will readily admit the entrance of the male organ when the two are brought together. The vaginal passage opens into, and terminates in the uterine, or womb cavity.

The womb is a pear shaped sack which is suspended in the womb cavity by cords and muscles from above. It hangs, neck downwards, and is, in its unimpregnated condition, about two and a half inches in diameter at its upper, or widest part, tapering to a thin neck at its lower end. It is hard and muscular in its quiescent state, filled with delicate and most sensitive nerves and capacious blood vessels. At its lower, or neck end, it opens directly into the vaginal passage.

The ovaries are two in number, and are situated on each side of, and above the womb, in the region of the upper groins. They are small, fan shaped glands, and are connected with the uterus by small ducts which are known as the fallopian tubes.

As already stated, the exterior parts of the body, in which the female sex organs are located, are covered with hair for their adornment and protection.

Such in brief, are the male and female sex organs in human beings. A further description of them and their functions and proper use we are now ready to consider.

V
The Function Of The Sex Organs

It hardly need be stated here, for it is a matter of common knowledge, that the *primary* purpose of sex in the human family is the reproduction of the race. In this respect, considered merely on its material, or animal side, mankind differs little from all other forms of animate life. As Whitman says, we see "everywhere sex, everywhere the urge of procreation." The flowers are possessed of this quality, and with them all vegetable forms. In the animal kingdom the same is true. Always "male and female" is everything created.

And the chief facts in reproduction are practically the same wherever the phenomena occur. Here, as everywhere else in the world, when a new life-form appears, it is always the result of the union of *two* forces, elements, germs or whatsoever. These two elements differ in nature and in function, and each is incomplete and worthless by itself. It is only by the combining of the two that any new result is obtained. It is this fact that has led to the most suggestive and beautiful phrase "The duality of all unity in nature."

Many centuries ago an old Latin philosopher wrote the now celebrated phrase, *Omne ex ovo*, which, translated, means *everything is from an egg*. This is practically true of all life-forms. Their beginning is always from an ovum, or egg. In this respect, the reproduction of human beings is the same as that of any other life-form.

Now in this process of producing a new life-form, the female is always the source of the egg, out of which the new creation is to come. This egg, however, is infertile of itself, and must be given life to, by mingling with its germ, an element which only the male can produce and supply. This element is technically known as a sperm, or a spermatozoa. Its function is to fertilize the dormant germ in the egg produced by the female, and thus to start a new and independent life-form. This life-form, thus started, grows according to the laws of its becoming more and more, until, at the expiration of a fixed period, which varies greatly in different animals, it becomes a complete young individual, of the nature and kind of its parents. The fertilization of the ovum in the female is called conception; its growing state is called gestation, and its birth, on becoming a separate being, is called parturition. In its growing condition, and before its birth, the new young life form is known as the foetus.

Now it is the fertilization of the ovum in the female (and from now on, it is only of the male and female in the human family that mention will be made) by the male, in the woman, by the man, that is of supreme interest and importance to both the parties concerned in producing this result. How this is brought about is substantially as follows:

As already stated, the infertile ovum, or egg, is produced by the woman. Such production begins at what is known as the age of puberty, or when the hair begins to grow upon the pubic parts of the female body. The time of the appearance of this phenomenon in feminine life varies from the age of nine or ten years to fifteen or sixteen. The average, for most girls, is fourteen years of age. At this time the formation of ova, or eggs, in the female body begins, and it continues, in most women, at regular intervals of once in twenty-eight days, except during pregnancy and lactation, for a period of about thirty years. During all this time, under favorable conditions, it is possible for the ovum produced by the woman to become fertilized, if it can meet the sperm of the male.

In a general way, this meeting of the infertile ovum of the woman with the sperm of the man can be brought about, as follows:

The ova are produced by the ovaries (the word ovaries means egg producers) where they slowly develop from cells which originate in these glands. When they have reached maturity, or are ready for fertilization, they pass out of the ovaries and down into the womb, by way of the fallopian tubes. As already stated, such passage of the ova from the ovaries into the womb occurs every twenty-eight days, and it is accomplished by a more or less copious flow of blood, a sort of hemorrhage, which carries the ova down through the fallopian tubes, and deposits them in the womb. This blood, after performing its mission of carrying the ova down into the womb, escapes from the body through the vaginal passage and is cared for by the wearing of a bandage between the thighs. This flow of blood continues for about five days, and is known as a menstrual flow; and this time in a woman's life is known as the menstrual period. It is so named because of the regularity of its recurrence, the word *mensa* meaning a *month*. In common parlance, these periods are often spoke of as the "monthlies."

After the ovum has reached the womb it remains there for a period of about ten days, after which, if it is not fertilized during that time, it passes out of the womb into the vaginal passage, and so out of the body. But if, at any time after it is ripe for fertilization, that is, from the time it begins its journey from the ovaries to the womb, and while it is in the womb, the ovum is met by the male sperm, it is *liable* to become fertilized — conception is possible. These are facts of the *utmost importance,* to be thoroughly understood and kept well in mind by all married people who would live happily together, as will be hereafter shown.

So much regarding the female part of the meeting of the ovum and the sperm. The male part of this mutual act is as follows:

The sperm, or spermatozoa, originate in the testicles. Each sperm is an individual entity and *several thousands* of them are produced and in readiness for use, *at each meeting* of the male and female generative organs; and if *any one* of the countless number comes in contact with the unfertilized ovum in the womb, conception is *liable* to result.

These sperms are so small that they are not visible to the naked eye, but they are readily seen by the use of a microscope. In shape they much resemble tad-poles in their earliest stages.

At the base of the penis, well up in the man's body, there is a large gland which surrounds the penis like a thick ring, and which is called the prostate gland. It secretes a mucous fluid which looks much like, and is about the consistency of the white of an egg. Close to this gland, and almost a part of it, is a sack, or pocket, into which the mucous secretion from the prostate gland is poured, and where it is kept, ready for use, in performing its part of the germinal act.

Now it is the business of this mucous fluid, which comes from the prostate gland, to form a "carrying medium" for the spermatozoa which originate in the testicles. There are small ducts leading from the testicles into the pocket which contains the prostate fluid. These are known as the seminal ducts, and through them the spermatozoa pass from the testicles into the prostate pocket. Here they mingle with the prostate fluid, in which they can move about freely, and by means of which they can be carried wherever this fluid goes. The combination of prostate fluid and spermatozoa is called "semen."

Seen under a microscope, a single drop of semen reveals a multitude of spermatozoa swimming about in the prostate-carrying medium. It is in this form that the vitalizing male element meets the female infertile ovum. This mass of live and moving germs is poured all around and about the region in which the ovum lies waiting to be fertilized, and every one of them seems to be "rushing about like mad" to find what it is sent to do, namely, to meet and fertilize the ovum. The manner of depositing the semen where it can come in contact with the ovum is as follows:

In order that this mingling of the male and female sources of life may be possible, it is necessary that there be a union of the male and the female generative organs. For such meeting, the penis is filled with blood, all its blood vessels being distended to their utmost capacity, till the organ becomes stout and hard, and several times its dormant size, as has been already told. In this condition it is able to penetrate, to its utmost depths, the vaginal passage of the female, which is of a nature to perfectly contain the male organ in this enlarged and rigid condition. Under such conditions, the penis is inserted into the widened and distended vaginal passage. Once together, a mutual back and forth, or partly in and out movement, of the organs is begun and carried on by the man and woman, which action further enlarges the parts and raises them to a still higher degree of tension and excitement. It is supposed by some that this frictional movement of the parts develops an electrical current, which increases in tension as the act is continued; and that it is the mission of the pubic hair, which is a non-conductor, to confine these currents to the parts in contact.

Now there are two other glands in these organs; one in the male and one in the female, which performs a most wonderful function in this part of the sexual act. These are the "glans penis" in the male and the "clitoris" in the female. The first is located at the apex of the male organ, and the other at the upper-middle and exterior part of the vulva. These glands are covered with a most delicate cuticle, and are filled with highly sensitive nerves. As the act progresses, these glands become more and more sensitized, and nervously surcharged, until, as a climax, they finally cause a sort of nervous explosion of the organs involved. This climax is called an "orgasm" in scientific language. Among most men and women it is spoken of as "spending."

On the part of the man, this orgasm causes the semen, which till this instant has remained in the prostate pocket, to be suddenly driven out of this place of deposit, and thrown in jets, and with spasmodic force, through the entire length of the penis, and, as it were, shot into the vaginal passage and the uterine cavity, till the whole region is literally deluged with the life-giving fluid. At the same time, the mouth of the womb opens wide; and into it pours, or rushes, this "father stuff," entirely surrounding and flooding the ovum, if it be in the womb. This is the climax of the sexual act, which is called "coitus," a word which means, going together.

With the myriads of spermatozoa swarming about it, if the vital part of the ovum comes in contact with some one of them, any one of which, brought into such contact, will fertilize it, conception results. The woman is then pregnant, and the period of gestation is begun.

This is a brief description of the act of coitus and of the means by which pregnancy takes place. It is, however, only a small part of the story of the sex relations of husbands and wives; and, be it said, a *very* small part of that, as will now be shown.

As has already been said, this use of the sex organs, merely to produce progeny, and so insure a continuance of the race, is a quality that mankind shares with all the rest of the animal kingdom. In all essentials, so far as the material parts of the act are concerned, the beginnings of the new life in the human family differ not a whit from that of any other mammals. In each case the ovum is produced by the ovaries of the female, passes into the womb, is there met by the semen from the male, fertilized by the spermatozoa, and so the foetus gets its start. This is the universal means by which the beginnings of all animal reproductive life takes place.

But there is another phase in the sex life of human beings, which is *entirely different* from that of all other animals, and which must therefore be considered beyond all that needs to be said regarding the act of coitus for reproductive purposes only. This we are now ready to consider and study.

Now in all animals, except human beings, the act of coitus is only permitted by the female, (it would seem is only *possible* for her) when the ovum is present in the womb and ready to be fertilized. *At all other times, all female animals, except woman, are practically sexless*. Their sexual organs are dormant, and *nothing can arouse them* to activity. Not only do they fail to show any desire for coitus, but if an attempt should be made to force it upon them, *they would resist it* to the utmost of their strength.

But when the ovum is present in the womb, these same female animals are beside themselves with desire for coitus. They are then spoken of as "in heat." And until they are satisfied, by meeting the male and procuring from him the vitalizing fluid which will fertilize their infertile ovum; or, failing in this, until the ovum passes away from them, out of the womb, they know no rest. At such times they will run all risks, incur all sorts of danger, do every possible thing to secure pregnancy. The thousand-and-one ways which female animals use to make known to their male mates their sexual desire and needs, when in heat, is a most interesting and wonderful story, a record made up of facts which would be well worth any student's knowing. But as all such knowledge can readily be procured from books which are within the reach of all, there is no need of noting the data here.

But now, *in woman, all these things are different!* As a matter of fact, the presence of the ovum in the womb of a normally made woman *makes little, and, in many cases, no difference whatever* as regards her status concerning the act of coitus! That is, women are never "in heat," in the same sense in which other female animals are. To be sure, in some cases, though they are rare, some women are conscious of a greater desire for coitus just after the ceasing of the menstrual flow; that is, when the ovum is in the womb. But such cases are so infrequent that they may well be counted atavistic, that is, of the nature of a tendency to return to a previous merely animal condition. For the most part, it is true of all normal women that the presence of the ovum in the womb makes little difference, one way or another, in regard to their desire for, or aversion to, the act of coitus.

Now the fact of this remarkable difference in the sex-status of women and the same quality in all other female animals leads us to a great number of interesting, not to say startling, conclusions, some of which are as follows:

In the first place, the phenomenon clearly establishes the fact that sex in the female human being *differs, pronouncedly,* from that of all other female life. For, whereas, among all females except woman, coitus is *impossible,* except at certain times and seasons, among women the act can not only be permitted, but is as much possible or *desired* at one time as any other, regardless of the presence or absence of the ovum in the womb. That is (and this point should be noted well by the reader) there is a *possibility,* on the part of the female humanity, for coitus, *under conditions that do not at all obtain in any other female animal life.*

This is a conclusion which is of such far-reaching importance that its limits are but dimly recognized, even in the clear thinking of most married people. The fact of such difference is known to them, and their practices in living conform to the conditions; but what it all means, they are entirely ignorant of, *and they never stop to think about it.*

And yet, *right here is the very center and core of the real success or failure of married life*! Around this fact are grouped all the troubles that come to husbands and wives. About it are gathered all the joys and unspeakable delights of the happily married — the only truly married. It is these items which make a knowledge of the real conditions which exist, regarding this part of married life, of such supreme importance. If these conditions could be rightly understood, and the actions of husbands and wives could be brought to conform to the laws which obtain under them, *the divorce courts would go out of business,* their occupation, like Othello's, would be "gone indeed."

The first conclusion, then, one that is forced upon the thoughtful mind by the fact of this difference in the sex possibilities of women and other female animals, is, as already stated, but which is here repeated for emphasis, that coitus *can* be engaged in *by women* when *pregnancy* is *not* its purpose, on her part; and that *this never occurs in any other form of female life!*

In view of this fact, is it too much to raise the question whether or not sex in woman is designed to fulfill any other purpose than that of the reproduction of the race? True it is, that the *only* function of sex in all other females is merely that of producing offspring — of perpetuating its kind. Under no circumstances does it *ever* serve *any* other end, fulfill any other design. *There is no possibility of its doing so!*

But one can help wondering if it is not true that, with the existence of the *possibility* of engaging in coitus *at will,* rather than at the bidding of *instinct* alone, there has also come a *new* and *added* function for the sex-natures that are capable of engaging in such before-unknown experiences? To a fair-minded person, such conclusion seems not only logical, but irresistible! That is in view of this conclusion, it naturally follows that sex in the human family is *positively designed to fulfill a function that is entirely unknown to all other forms of animal life.* And from this, it is but a step to the establishment of the fact that *sex exercise in the human family serves a purpose other than that of reproduction!*

Now, this fact established, a whole world of new issues arises and demands settlement. Among these, comes the supreme question: *What is the nature of this new experience that has been conferred upon human beings, over and above what is vouchsafed to any other form of animal life? What purpose can it serve? How can it be properly exercised? What is right and what is wrong under these new possibilities*? These are some of the issues that *force* themselves upon all thoughtful people, *those who wish to do right under any and all circumstances in which they are placed.*

Of course, here as elsewhere, the unthinking, the happy-go-lucky and the "don't-give-a-damn," can blunder along in almost any-old-way. But they can, and will, reap only the reward which always follows blundering and ignorance. In these days of scientific clear-thinking, we have come to understand that *salvation from sin comes by the way of positive knowledge and not at the hands of either ignorance or innocence*! If husbands and wives ever attain to the highest conditions of married life, it can only be after they *know and practice, what is right in all their sex relations, both for reproductive purposes and in all other respects! Note that well*!

As things are now, especially in all civilized countries, and particularly among Christian people, this *secondary* function of sex in the human family, while blindly recognized as a fact, is none the less abused, to a most shameful degree. For ages, the whole situation has been left in a condition of most deplorable, not to say damnable, ignorance; and no honest endeavor has been made to find out and act up to the truth in the premises. Husbands and wives have engaged in coitus *ad libitum*, utterly regardless of whether it was right or wrong for them to do so! They have taken it for granted that *marriage* conferred on them the *right* to have sexual intercourse whenever they chose, (especially when the man chose,) and they have acted accordingly. This is especially true of men, and the practice has been carried to such length that the right of a man to engage in coitus with his wife *has been established by law,* and the wife who refuses to yield this "right" to her husband can be divorced by him, if she persists in such way of living! It is such a fact as this which caused Mr. Bernard Shaw to write: "Marriage is the most licentious institution in all the world." And he might rightfully have added "it is also the most brutal," though it is an insult to the brute to say it that way, for brutes are never guilty of *coitus under compulsion. But a husband can force his wife to submit to his sexual embraces, and she has no legal right to say him nay!* This doesn't seem quite right, does it?

Now there are several different ways of viewing this new and added sexual possibility in the human family, namely, the act of coitus for other than reproductive purposes. The Catholic church has *always* counted it as a sin. Popes have issued edicts regarding it, and conclaves of Bishops have discussed it and passed resolutions regarding it. There has always been a difference of opinion upon the subject amongst these dignitaries and authorities, but they all agree in one respect, namely, that it is a *sin.* The only point of difference has been as to the extent or enormity of the sin! By some it has been reckoned as a "deadly sin," punishable by eternal hell fire, if not duly absolved before death; by others it has been held to be only a "venial sin," one that must always be confessed to the priest when coitus is engaged in, and which can be pardoned by the practice of due penance. *But, always, it was a sin!*

The Protestant church has never issued edicts regarding this matter, but, for the most part, it has tacitly held to the Catholic teaching in *theory*, while universally *practicing* the reverse, in actual married life. Protestants have looked upon it as a necessity, but have taught that it was *regrettable* that such was the case. They have held, with Paul, that, "it is better to marry than to burn." And most of them have chosen the marriage horn of the dilemma.

Among some European nations, attempts have been made to make it impossible for husbands and wives to cohabit except for reproductive purposes. In one of these nations, padlocks were used for preventing the act. A slit was made through the foreskin of the penis, and through this slit the ring of a padlock was passed, much as an ear-ring is passed through the lobe of a lady's ear. The padlock was made so large that it could not be introduced into the vaginal passage, and so coitus was impossible when it was worn. It could only be removed by the magistrate into whose hands the regulation of this part of the citizens' life was given. Specimens of these padlocks are still to be seen in European museums.

Now the terribly immoral thing in all this way of living has always been the fact that it *compelled* people to continually *violate their consciences*, by *pretending* to *believe* one thing and constantly *practicing* the reverse of their proclaimed belief. That is, it lured them into *living a continual lie, and such can never be for the good of the soul*! It goes without saying that the sooner this abominable way of living can be ended, the better it will be for all parties concerned — the individuals who are the victims of such falsehood, and the communities of which they form a part.

From all this it follows that the first thing every

new husband and wife *ought* to do is to *settle clearly in their own minds the issue as to whether it is right or wrong for them to engage in coitus for any other than procreative purposes.* Having settled this point, one way or the other, then *let them conscientiously act accordingly. For only so can they live righteous lives*!

In settling this point, so far as available authorities for the young people to study and consider are concerned, these are all *against* coitus except for begetting of off-spring. All the "purity" writers and Purity Societies are ranged together on the negative side. Likewise are all the books of "advice to young wives and husbands," especially those addressed to young *wives*.

Now all these "authorities" base their whole argument upon the purely *animal* facts in the premises. Probably a certain Dr. C. is more largely read for information on these matters than any other author, especially among young women. He has written a large, and from the view-point he takes, a very plausible volume; and it is very extensively advertised, especially in papers which young women read. The result is that it has come to be almost a standard authority in these affairs.

Dr. C.'s argument is, baldly, as follows: — (a) Among animals, the universal practice is a single act of coitus for each begetting of off-spring, (b) Human beings are animals, (c) Therefore, human beings should only engage in coitus for reproductive purposes.

To this syllogism he adds a corollary, which is, that, therefore, all sexual commerce in the human family, for any other than reproductive purposes, is *wrong*. These are his texts, so to speak, and through several hundred pages he preaches, *don't, don't, don't,* sermons. The entire volume is one of denial and prohibition. He proclaims the act, even for the one purpose he allows to be right, as low, and in itself degrading, to be engaged in only after "prayer and fasting" and "mortifying the flesh," and even then, in the most passionless, and only done-because-it-has-to-be manner; as a mere matter of duty; to be permitted by sufferance; joyless, disgusting in itself; a something to be avoided, even in thought, other than it is a necessity for the continuance of the race.

It is from such data as this that thousands of "innocent" brides annually make up their minds as to what is right or wrong in the matter of sexual intercourse.

In doing this, most of these young women are perfectly conscientious, and want to do the right thing, and there are two items in the count that naturally lead them to accept Dr. C.'s teachings as correct. The first is, that it coincides with all they have ever heard about such matters; the second, that the Doctor flavors all his text with a religious quality, of the alleged most sacred sort. He instances saintly women who have lived the most ascetic lives, and whose religious status was achieved because, and by means of, their perfect chastity. In fact, this word "chastity" (which he translates as entire renunciation of the whole sex nature) becomes the test word of his whole treatise, and its practice is upheld as the true road to all goodness and virtue.

Now, nearly all well-bred and cultivated young women are naturally religious (and not a word should be said against their being so) and they are anxious to time their lives to everything that the highest religious demands prescribe. It is, therefore, most natural that, being thus taught by an authority for which they have the highest regard, they enter marriage with the *fixed opinion* in accordance with their teaching. How could it be otherwise?

On the other hand, a few young husbands, indeed none but now and then a "goody-good" (who usually turns out to be the worst of the whole lot, in course of time), are willing to "stand for" any such theory, much less to live any such life as this theory would impose. These "don't care what the book says," and, from the manner of their bringing up, from all they have learned or heard by hearing *men* talk about married life, (which is usually of the most vulgar sort) they have come to the conclusion that marriage confers upon the parties the *right* to engage in sexual commerce at will; and, especially, that the husband has the *right* to the body of his wife *whenever he chooses*. For, indeed, does not the law give him that right! And so long as one "keeps inside the law" what more could be asked! Yea, verily! What more could be asked?

So it is that *most brides and bridegrooms go to their marriage bed with the most widely diverse views as to what is right and wrong in the premises* — as to the life they will lead in their new estate. The young wife is for "purity" and "chastity." The young husband, driven by a passion which he has long held in thrall, in the belief that he can now give the fullest vent to it, when he has got where such relief is possible, is like an excited hound when it seizes its prey, which he fully believes he has the right to deal with as he pleases! What wonder that, in view of all these circumstances, the most extensive observer of marriage-bed phenomena should write: "*As a matter of fact, nine young husbands in ten practically rape their brides at their first sexual meeting.*" *Could anything be more horrible, or criminally wicked*? And it is all so needless! It is all the result of ignorance, of "innocence," and the worst of false teaching. The pity of it!

True, these unfortunate conditions are often modified by "mother nature," who inspires the bride with curiosity, which, in a measure, controls her in spite of her false teachings, and with passion, which, to a degree, will assert itself over and above all false modesty, her religious scruples and her fear of pregnancy; and so she *may* come through the ordeal of introduction to the act of coitus in a fairly sane condition of mind, even though she may have practically been *raped*! But, too often, the result of such first contact is *a shock to the bride from which she may not recover during all the subsequent years of married life*! And "here is where the trouble lies," for untold thousands of married men and women, all over the civilized world, to-day. And it might all be so different! It ought, *in every case*, to be all so different! But if it ever does become different, *knowledge* has got to take the place of "*innocence*" on the part of the *bride*, and of *ignorance* on the part of the *bridegroom*, both of whom must be *taught* to "*Know what they are about*" before they engage in the sexual act, and be able to meet each other sanely, *righteously, lovingly,* because they both *desire* what each has to give to the other; in a way in which neither claims any *rights*, or makes any *demands* of the other – in a word, in *perfect concord* of agreement and action, of which mutual love is the inspirer, and *definite knowledge* the directive agent.

Such a first meeting of bride and bridegroom will be no raping affair. There will be no shock in it, no dread, no shame or thought of shame; but as perfectly as two drops of water flow together and become one, the bodies and souls of the parties to the act will mingle in a unity the most perfect and blissful that can ever be experienced by human beings in this world. This is no dream! It is a most blessed reality, which all normally made husbands and wives can attain to, if only they are properly *taught and educated,* if only they will learn how to reach such blissful condition.

However, such greatly desired status is not to be had for the asking merely. *Instinct can never bring it about; "innocence"* will never yield such a result; and *force,* or the declaration of a "*right*" in the premises will forever banish it to the realm of the never-to-be-realized. It can only come as a result of clear-headed thinking, scientific investigation, honest study, wise and righteous action under the given conditions; and, above all, *a love, each for the other, that knows no bounds.* All these things *must* obtain, *on the part of both parties concerned,* or the desired results can *never* be attained.

Having said which, here shall follow some suggestions as to how such estate may be reached by the readers of these pages.

But first, let us finish Dr. C., and all of his tribe — banish them from all our reckoning in these matters, forever.

As already shown, this argument has not a leg to stand on. These writers treat the whole situation as though men and women were *mere animals! Men and women are far more than mere animals, and God hath made them so*! And for these reasons we will have respect for men and women as *God has made them,* rather than as Dr. C. and the "purity leagues" say God *should* have made them!

As a matter of fact, the secondary function of sex in the human family is something *far above* mere animality; it is something that mere animals know nothing about, that they can never experience, or in any way attain to, and these *fundamental differences* in the premises remove the whole issue from the realm of comparison with any forms or functions of mere animal life. As well reason that animals never eat cooked food, and so men ought never to eat cooked food (and there are some people who do so reason, strange to say) or that animals do not wear clothes, and so men ought not to wear clothes — as well make these, or a score more of comparisons, between the human race and mere animals, as to try to compare them in the item of their sex functions.

In only the single fact that, on the physical plane merely, coitus for the purpose of procreation is common to all animal life, mankind included, is there a point of comparison between humanity and the brute creation. *Beyond that point there is nothing comparable between the two*! As well say that because beasts can hear, therefore they can comprehend and enjoy a Beethoven Sonata, or because they have eyes they can delight in a picture by Corot!

This is only another way of saying that sex has functions and uses in the human family that are entirely apart from the possibilities of all other animal life — functions as much above mere animality as music is above mere physical hearing, as painting above mere physical sight.

These facts forever upset and overthrow all the theories of Dr. C. and Co., they entirely eliminate the whole bunch from any part or lot in the issue on which they have essayed to speak with such authority, but whose main point, whose essential elements they have *entirely misunderstood*, and hence have treated in a way that is wholly at variance with the truth in the premises, and it is the truth that we are looking for.

Once more (for it is well to go to the bottom of this matter while we are about it) the honest truth is, that *it is the universal practice of the human race for men and women to cohabit for other purposes than reproduction, and it has always been so,* since men and women were men and women! It is true among the most savage and barbarous tribes of the earth, and it is more emphatically true of the highly civilized people in all lands and climes. And is it reasonable to suppose that such a universal phenomenon should *not* have been intended to be as it is! As well say that appetite for food is a mistake, one that ought to be eliminated!

Again, the experiences of men and women, all over the world, prove that, where this act is engaged in properly, according to the laws that obtain in the premises, *it conduces to the highest physical, mental, and spiritual well-being of the parties concerned.* Indeed, it is beyond doubt true that the men and women who have never known this most perfect of all human experiences, have never reached the summit of human attainment, have never arrived at the perfection of manhood and womanhood. Length of life, health of the highest sort, and happiness, the most delectable — all come, these and more, to men and women by this route, *if it is rightly traveled.* Hell and damnation result if that road is wrongly trod!

And that's what makes the manner of traveling it so important.

VI
THE ACT OF COITUS

Strictly speaking, the act of coitus should be considered as composed of four parts, or acts, of one common play, or drama. Not that there is a sharp line of demarcation between each act or part, for the *four* really blend into *one* composite whole, when taken together, seriatim; but there are *four phases* of the act which may well be studied separately, in making a detailed review of a sexual meeting of a man and a woman.

These four parts are: *first,* the preparation for the act; *second,* the *union* of the organs; *third,* the motion of the organs; *fourth,* the orgasm.

In what immediately follows, these *four* stages of the act of coitus will be studied and traced in detail, with the utmost care, in the hope that such pursuit may result in the best possible good to the student.

Regarding the *first* part of the act, let it be said that here, above all other situations in the world "*haste makes waste.*" *Put that down as the most fundamental fact in this whole affair!* Right here is where ninety-nine one-hundredths of all the troubles of married life begin! And the fault, right here, is usually (though not always) with the husband! But he doesn't mean to be bad. Not once in a thousand times does he deliberately purpose to do wrong. He is simply the victim of undirected and ungoverned passion, and of an *ignorance* which results in stupid blundering, or carelessness, or thoughtlessness. What such a husband practically does is to rush blindly and furiously along a way he knows nothing of, but which he has been led to think he has a *right* to travel *when and how he will*! The ordinary figure of a "bull in a china shop" can but faintly describe the smashing and grinding to powder of the most delicate situation that can occur in all human experiences, that result from such action as this. Ideals that have touched heaven are tumbled from their lofty places and ruthlessly crushed to atoms; hopes that were beyond the power of words to express go out in despair; dreams become a hideous nightmare; and love, which was as pure as crystal waters, is muddied, befouled, and made into a cesspool! *And all this because of ignorance* or careless hurrying, of making haste where the utmost of time, caution and intelligent care should have obtained!

As has already been explained, when the act of coitus is to be engaged in, the sex organs of both the man and the woman undergo great changes. Blood rushes to all these parts, in copious quantities, till they become gorged. The result is that the penis is enlarged to several times its dormant size, and the vulva and vagina should, and will, under right conditions, undergo similar changes and transformation.

But there is usually a great difference in the length of time it takes for these changes to take place in men and women. On the part of the man, as soon as his passion is aroused to any considerable extent, the penis at once makes itself ready for action. It "tumesces," or swells itself hard, almost instantly; and, so far as its mere physical stoutness is concerned, is as ready to enter the vagina then as ever, even if it has to force itself in.

On the other hand, the tumescence of the parts in women is usually, (especially as girls are reared) not infrequently, a matter of considerable time, not infrequently several minutes, and now and then, of *half-an-hour or more*! This is not always so, for in some very passionate women they are ready for action almost instantly. Indeed, there are some women whose sex organs tumesce if they (the women) even touch a man — any man — and occasionally a case occurs where a woman will experience an orgasm if her clothing brushes against a man! Such cases are, of course, abnormal. But, *for the most part*, it is true that women are *much slower* in making ready for the sexual act than men are.

Again, as the organs become ready for the act, nature has provided a most wonderful means for bringing about their easy and happy union. Both the male and female organs secrete and emit, or pour out, a sort of lubricating fluid which covers and sometimes almost floods the parts. This is a clear and limpid substance, that looks much like the white of an egg, and is much like the saliva that is secreted in the mouth, only it is a thicker substance. Chemically, it is almost identical with saliva. That generated by the man is called "prostatic flow;" that produced by the woman "pre-coital secretion."

Now, if time is given for this fluid to be secreted and exuded, all the parts become covered or saturated with it, and they are admirably equipped for easy union. The glans penis is then covered with the slippery fluid, and the vulva and all the walls of the vagina are laved with the substance. At the same time, the vaginal walls have widened and grown soft, and all the parts of the vulva (which are yet to be named and described in detail) are in like condition. The result is that, though the penis be what might at first seem of such size as to make its entrance into the vagina impossible, as a matter of fact, such entrance is perfectly easy, when the parts are fully ready to be joined. *But not before or otherwise!*

So here is where the trouble comes. If the husband is in haste, if he does not wait for the wife to become ready to meet him; if he forces his large, hard penis into the vagina before either is fully ready for such union — when there is no prostatic fluid on its glans, and the vagina is shrunken and its walls are dry — if coitus is engaged in in this way, it is perfectly easy to see that *only disaster can result*! The woman is hurt, sometimes most cruelly, and the man in reality gets only a beastly gratification from the act. *Of all bad things in all the world, such manner of coition is the worst*!

And so, in this *first* part of the act, the one foremost thought to remember and observe is, *take plenty of time!*

There is another reason why, on the part of woman, this time should be extended, especially when she is a bride and inexperienced in these matters, and that is, that her "innocence," and all her education, make her feel that she is *doing wrong*, or at least permitting a wrong thing to be done, and this holds back the proper growth of her passion, hinders the tumescence of her sex organs, delays the flow of the precoital secretion, and so keeps her from becoming properly prepared for her share of the mutual act.

Again, her fear of pregnancy may still further retard her coming into a proper condition. Indeed, this last is the almost common cause for her failing to be in readiness for meeting her husband. All of which items must be taken into account by both husband and wife, and intelligently, lovingly dealt with, if the best results for both parties are attained.

As regards the item of possible pregnancy, special note will be made of this feature later on. It is here placed in abeyance for the time being, because its consideration can be better provided for after some other points have been studied.

Now the one easily understood (and as easily practiced as understood) direction as to what to do by way of preparation for the act of coitus is: *do as lovers do when they are "courting."* And everybody knows what that is! And note this — that *nobody ever hurries when they are courting!* They delay, they protract, they dilly-dally, they "fool around," they pet each other in all sorts of possible and impossible ways. They kiss each other — "long and passionate kisses, they again and again give and receive" — they hug each other, nestle into each other's arms — in a word, they "play together" in a thousand-and-one ways which the "goody-goods" declare to be wrong, and the cold-blooded call nonsense or foolishness, but which all *lovers* know is an *unspeakable delight* ("unspeakable" is the word, for who wants to *talk* when these blissful experiences are going on!).

Now, these things, and the likes of these things, in limitless supply, should always precede the act of coitus. It is right there that this part of the first act of this wonderful four-act drama or play should be wrought out, and if they are omitted or disregarded, the play will end in *tragedy, with all the leading actors left dead upon the stage*!

Now the chief, if not the only, reason why this part of the supreme act of married life is not always preluded in this way is found in the *false view* of what the *marriage ceremony means*, and a wrong impression as to what it confers upon the parties who say "yes" to its prescriptions. That is, the common idea is, that the taking of "marriage vows" bestows certain *rights* and imposes certain *duties* upon the new husband and wife. It is thought that such ceremony makes certain acts *right* which would *otherwise* be *wrong*, and that it establishes the *right* to engage in such acts, *with or without any further consultation or consent in the premises*. It makes love a matter of *contract*, a something *bound by promise and pledge rather than a free and unfettered effusion of the soul*.

The result of this is that, whereas, before the marriage ceremony both the man and woman take the utmost care to do everything in their power to increase, magnify, and retain each other's love, after they have been granted a "license," and the minister has put their hands together and prayed over them — after this, they both think they have a "*cinch*" on each other, that they are bound together by a bond that cannot be broken, a tie so strong that it will need no further looking after, but which will "stay put" of its own accord, and which may therefore be let to shift for itself from the hour of its pronouncement! Nothing *could be further from the truth than this is*. And yet it is a common feeling and belief among young married people!

Nor is it any wonder that this should be so. The very form of the marriage ceremony and contract tends to make it so. The fact that marriage originated as a form of slavery, and that much of its original status yet remains — all these things tend to establish these wrong ideas regarding the estate, in the minds of the parties to it.

Nor are the evils that come from such wrong view of marriage all confined to one side of the house. On the contrary, they are about evenly divided between husbands and wives, witness a few illustrations, as follows:

A couple had been married about a year. They

had no children, nor were there prospects of any. The husband was beginning to spend his evenings away from home, leaving his wife alone. One evening, as he was making ready to go out, his wife said: "What makes you go out evenings now, and leave me alone! You didn't use to do it!" And the husband replied:

"Why, you don't do anything to make it interesting for me now! You used to put on your prettiest clothes when I came to see you, fixed up your hair bewitchingly, had a smile for me that wouldn't come off, would sing for me, read to me, sit on my lap and pet me and kiss me, and now you never do anything of the kind." And before he could say more, the wife responded: *"Oh, but we are married now, and it's your duty to stay with me!"*

What wonder that the husband went out of the house, slamming the door after him! The wonder is that he ever came back.

Again: A woman who was a graduate of a famous Eastern College and who had taught for a number of years, who was from one of the "first families" in the east, and was counted as a lady of the highest culture and refinement, finally married a Western business man. On their bridal night, as they were retiring, the man laid his hand on the woman's bare shoulder, and she threw it off, and said: "Don't be disgusting! I married you because I was tired of taking care of myself, or of having my relatives take care of me. You are worth fifty thousand dollars, and one-third of all that was made mine just as soon as the preacher got through his closing prayer, and you can't help it! That's the truth, and we are married, and you can make the best of it!"

These are both truthful tales, nor are they the

only ones of the sort that could be told.

On the other hand, these are matched with acts of ignorant and careless young husbands, who do dastardly deeds to their brides because they think *the law* and the *contract* give them the right! There is no need to go into details. The whole evil is revealed by the words of the woman just quoted: "*Oh, but we are married now.*"

These records, and all like them, lead to the remark that *marriage confers no rights, to either the bride or the bridegroom, in the highest meaning of the word.* So far as its outward and formal observance is concerned, marriage is merely a sort of protection for society which has grown up through the years, and which is probably for the best, for the present, things being as they are. But it should be well understood that it can *never* lead to *true happiness* if it is viewed and utilized *merely* on its *legal and formal side. True marriage is based on mutual love; and mutual love can never be traded upon, or made an item of formal agreement and contract.* People may contract to live together and to cohabit, and they may faithfully carry out their agreements; *but this is not marriage*! It is simply *legalized prostitution, bargain and sale, for a consideration. It is blasphemy to call it by the sacred name of marriage!*

Truly does Tennyson say: "Free love will not be bound." Indeed it cannot be! It must remain forever free if it stays at all. And if the parties to it try to bind it, the more chains, fastenings, pledges and agreements they put upon it, the sooner and quicker will it escape from all its holdings and fly away and *stay away*!

And so, to come back to where we left off (for we said there should be no hurrying or haste here) let married people understand that the key to married happiness is *to keep on "courting" each other*. Indeed, to make courting continually grow to more and more. During the whole extent of married life, never neglect, much less forget to be lovers, and to show, *by all your acts,* that you are lovers, and great shall be your reward. Don't ask how to do this! You know how, well enough. Do it!

And be careful *not* to do anything that a careful lover ought not to do! This direction should be heeded by both husband and wife. Make yourself beautiful for your husband, Oh, wife, and keep yourself so. As between the public, or your friends, or society, give them what of yourself you can spare, after you have given to your lover all that you can bestow upon him, or he can wish you to bestow. Don't give to everybody and everything else, church, society, work, children, friends, or what-so-ever — don't give *all* of yourself to these, and let your husband "take what there is left." Don't do that, as you value your married success and happiness! Don't say: "Oh, but we are married now," and let it go at that!

The beautiful and delicate flowers of married love need to be watched and tended with the most skilful care, *continually,* by both husband and wife. Treated in this way, they will not only be fragrant and lovely through all the years of wedded life; but as, one by one, the blossoms shed their petals and change their forms so that luscious fruits may come in turn — as these changes take place, new, more beautiful and more fragrant flowers will continue to the very end of the longest married life. Don't ever forget this, or doubt it, as you hope for happiness in the marriage state! Mind what is here said, and act accordingly *all the time* — days, nights and Sundays.

Now if these truths are thoroughly inculcated, "kicked in" so firmly and deeply that they will never "jar loose" or get away, we will move on.

So, then, the *first* part of *every* act of coitus should always be a *courting* act, in which there should be *no haste*, but in which the parties should "*make delays*," as John Burroughs says.

And this should be added: that, for married lovers, courting has a far wider range of possibilities than it has for the unmarried. Previous to marriage, there are conventionalities and clothes in the way! After that, neither of these need be in evidence, and this makes a lot of difference, and all in favor of the best results, if rightly used, and made the most of. One hardly need to go into details here, (though this may be done later on in this writing). If the lovers will be as free with each other unclothed as clothed; if they will utterly ignore all conventionalities, and do with and for each other anything and everything that their *impulses* and *inclinations* suggest, or their desires prompt; if they will, *with the utmost abandon* give themselves up to petting each other in every possible way that *mother nature* has put within their reach; if they will hug and kiss and "spoon" and "play with each other" just as they want to do – if they will do this, and not *hurry* about it – then, in due course, they will successfully execute the *first act* of the great play they are performing; the sex organs will become fully ready for the union they are both longing for; the "prostate flow" will have added to the erect condition of the penis; the walls of the vagina and all the area of the vulva will be enlarged, soft, flexible and made smooth and slippery by a most generous supply of the "pre-coital secretion" and everything will be in *perfect readiness* for the next part of the performance, namely the union of the organs.

And here it becomes necessary to say something about the position of the parties in making such union. There are a large number of these possible, some of which may be noted later, but here, only the most common one will be considered (it is said there are more than forty different positions possible in this act).

The most common position is for the woman to lie flat on her back, with her legs spread wide apart, and her knees drawn up so that the angle made by the upper and lower part of the leg shall be less than a right angle. Her head should not be too high, there should be no pillow under it.

Into her arms, and between her spread legs as she lies thus, her lover should come. His body will thus be over and above her, and *he should sustain himself on his elbows and knees*, so that little or *none* of his weight may rest upon her. In this position, face to face (and it should be noted that only in the human family is this position of coitus possible! Among mere animals, the male is always upon the back of the female. They — mere animals — can never look each other in the eye and kiss each other during the act! This is another marked and very significant difference between human beings and all other animals in this regard) it is perfectly natural and easy for the organs to go together, when properly made ready, as here-before described. The woman should also place her heels in the knee-hollows of her lover's legs, and clasp his body with her arms.

The entrance of the penis into the vagina should not be too abrupt, unless circumstances are perfectly favorable for such meeting and it is *the wish of the wife* that it be made in this way. It is only fair to say, though, that such bold and pronounced entrance is often *greatly desired by the woman,* if her passion has been fully aroused at this stage of the act. Such union is not infrequently of the greatest delight to her, if everything is favorable for its being so made. But, if there is any pain produced in her by the coming together, the meeting should be gentle and slow, the penis working its way into the vagina by degrees, till, finally, it is entirely encased therein. Once thus happily together, the vagina and uterine cavity will still further expand, till, in due order, the two organs will be fitted together perfectly, a single unit, *one,* in the highest sense of unity.

This is the *second* act in this wonderful play.

Once well together, and the organs perfectly settled and adapted to each other, the *third* act begins, namely, *the motion of the organs* – the sliding of the penis back and forth, partly in and out of the vagina, though this is not really the best way of describing just what should take place. What *should* actually be done is, that the *two* organs should engage in this motion, which is *common to them both.* They should *mutually* slip a few inches, back and forth, *each party to the motion doing a fair half.*

It is often supposed, by both an uninitiated husband and an "innocent" wife, that all the motion should originate with the husband – that he should slide his penis in and out of the vagina, while the woman should lie still and *"let him do it all."* This is, however, a *great* mistake, and one that has caused an endless amount of ill to untold numbers of husbands and wives. And for the following reasons:

In the position just described, if the wife has her

arms around her lover's body and her heels in his knee-pockets, while he supports himself by his elbows and knees over and above her, resting *none* of his weight upon her, it is perfectly easy for her to lift her hips up and down, or sway them from side to side, or swing them in a circling "round-and-round" motion, as she may choose to do. She can thus *originate* her half of the in-and-out motion — a something she will delight to do, *if given a fair chance.* If, however, the man lies heavily upon her, holding her down with the weight of his body, the possibility of such action on her part is prevented, and this results disastrously to both parties. And so, in this part of the act, the husband should take the *utmost care* to give his wife the *full and complete freedom* to move her hips as she chooses, and as a successful climax demands that she should.

Now if the wife be left free to move, as just described, and the in-and-out motion proceeds as it should, what immediately follows will vary in a great degree. Thus, the time taken to reach the climax, or last act of the performance, may be a few seconds, or several minutes, may require a mere half dozen motions, or *several hundred!* All depends on the intensity of the passions of the husband and wife, especially the latter, and their skill in manipulating this part of the act.

The effect of this motion is to still further excite and still more distend all the organs involved. Normally, the motion grows faster and faster, the strokes becoming as long as the length of the organs will possibly permit without separating them. The flow of the lubricating fluids, from both organs, becomes more and more copious, till, all at once, the orgasm, or *fourth stage,* is reached!

It is difficult to describe what this orgasm is like. There is no bodily sensation that at all corresponds to it, unless it be a sneeze, and this is only like it in that it is spontaneous, and a sort of nervous spasm (a sneeze is sometimes spoken of as an orgasm). A sexual orgasm is a nervous spasm, or a series of pulsating nervous explosions which defy description. The action is entirely beyond the control of the will, when it finally arrives, and the sensation it produces is delectable beyond telling. It is the topmost pinnacle of all human experiences. For a husband and wife to reach this climax, at exactly the same instant, is a consummation that can never be excelled in human life. It is a goal worthy the endeavor of all husbands and wives, to attain to this supreme height of sexual possibilities.

On the part of the man, the orgasm throws the semen into, and all about the vaginal-uterine tract. The amount of semen thus discharged at a single climax is about a tablespoonful, enough to entirely flush and flood the area into which it is thrown. Its use and action there have already been described, and so need not be repeated here.

On the part of the woman, the orgasm causes no corresponding emission of fluid, of any sort, that is jetted forth as is the semen. Yet the spasmodic action of the sexual parts, so far as nervous explosions are concerned, is exactly like that of her partner. Palpitation follows palpitation, through all the sexual area; the mouth of the womb opens and closes convulsively, the vagina dilates and contracts again and again, and the vulva undergoes similar actions. The sensations are all of the most delectable nature, the whole of the woman's body being thrilled, over and over, again and again, with delights inexpressible. This, however, seems to be the entire mission of the orgasm in woman. *It has nothing whatever to do with conception;* though many people, especially young husbands who know just a little about the phenomenon, believe that it is an *essential* to pregnancy. *But such is by no means the case.* All that is needed to bring about conception in a woman is the presence of the ovum in the uterus, and its meeting semen there, and so becoming fertilized. So far as becoming pregnant is concerned, the *woman* need have *no pleasure at all* in the act of coitus. Indeed, women have been made pregnant by securing fresh semen from some man and injecting it into the vagina with an ordinary female syringe!

The false idea, which largely prevails, and which usually takes the form that there is no danger or possibility of conception unless the orgasm is *simultaneous on the part of the man and woman,* has caused many a woman to become pregnant when she thought such a result to be impossible, because she and her lover did not "spend" at the same instant. For the same reason, many a young husband has impregnated his wife when he least expected to do so, thinking that because he alone experienced the orgasm, that therefore conception was impossible.

Again, there are many married men and women who do not know that it is possible for a woman to experience an orgasm at all! The writer once knew a case of this kind, where a husband and wife, most intelligent and well cultivated people, lived together for twenty years, and to whom were born six children, who, at the end of that time were wholly unaware of such possibility! They afterwards discovered it by accident, as it were, and after that enjoyed its delights for many years. There are some, yea, many, women who never experience this sensation at all, but of this more will be said later.

All these phenomena seem to indicate the fact that, so far as women are concerned, *the orgasm is entirely for her delectation and delight. It forms no part of the act of conception*, and its only possible function, beyond that of pleasure, is that, because of the exceedingly delightful sensations it produces, it may lure women to engage in coitus when, but for this fact, they would not do so, and that it thus increases the possibility of women becoming mothers. Indeed, there is no stronger temptation to a woman to run the risk of becoming pregnant than her desire to experience an orgasm! But more of this later.

As soon as the orgasm is over, a total collapse of the husband and wife takes place. They are truly "spent," a most expressive word, which alone can describe their condition. On the part of the man the up-to-this-moment stout penis, becomes almost instantly limp and shrunken, while all the female organs become quiescent. A most delightful languor steals over them; every nerve and fibre of the whole body relaxes; and a desire to fall asleep at once, comes upon them irresistibly. And the thing for them to do is to avail themselves of such natural impulse, just as soon as possible. They should always have at hand, and within easy reach, a towel, or napkin, with which to care for the surplus of the seminal emission, which, as soon as the organs are separated, will, in greater or less quantity, flow from the vagina. Some of the same fluid will also remain upon the penis when it is withdrawn. The husband should absorb this surplus which remains with him with the towel, as soon as the organs are parted, and immediately leave his superimposed position, leaving his wife *perfectly free*, to do as she will. She should arrange the towel between her thighs, exactly as she would a sanitary napkin, making no attempt to remove the surplus semen at that time, and turn over and go to sleep *immediately.* (It is said that if the woman goes to sleep on her *back,* after coition, she thereby increases the *probability,* of becoming pregnant. This is a point that women who greatly desire motherhood should note. The writer knew one case where a wife lay on her back for twenty-four hours after coition and so became pregnant after all other means had failed.)

Now it might seem that such neglect, on the part of the woman, to immediately remove the surplus semen, was uncleanly and unsanitary. But this is not at all true, and for this reason: *The semen is a most powerful stimulant to all the female sex-organs, and to the whole body of the woman.* The organs themselves will absorb quantities of semen, if left in contact with it, and it is most healthful and beneficial to them, and to the woman, to have them do so. It is for this cause that many women increase in flesh, and even grow fat after they are married and so can avail themselves of this *healthful food.* As a matter of fact, *there is no nerve-stimulant, or nerve-quieter, that is as potent to womankind as semen.* There are multitudes of "nervous" women, hysterical even, who are restored to health, and kept in good health, through the stimulative effects of satisfactory coitus and the absorption of semen, when both these items are present in perfection. On the other hand, there are many women who suffer all sorts of ills, when these normally beneficial factors are misused or wrongly applied. The results that follow all depend upon the way the act is done, and its products utilized.

So, after the act of coition is over, let the woman slip a "bandage" into place as soon as possible, and go to sleep. If she sleeps long, so much the better, so much more will she be benefited by the presence of the semen and its absorption. When she naturally wakens, she may bathe the vulva region with warm water; but there is no need of, nor is it wise to try to cleanse the vagina and the uterine tract by the use of a vaginal syringe. Above all, never inject cold water into the vagina, especially do not do this immediately after coitus. Some women use a cold water injection immediately after coitus. There is no surer way to ill health and ultimate suicide. The parts are congested with blood at such times, and to pour *cold* water upon them is as though, when one is dripping with perspiration, he should plunge into a cold bath. Nature has made wise provision for taking care of all the semen that remains in the vagina. Let the parts alone, and they will cleanse and care for themselves.

Such, then, is a somewhat extended review of the act of coitus at its best estate, and in a general way. *Its perfect accomplishment is an art to be cultivated, and one in which expertness can only be attained by wise observation, careful study of all the factors involved, and a loving adaptation of the bodies, minds and souls of both the parties to the act. It is no mere animal function.* It is a *union,* a *unity* of "two *souls* with but a single thought, two hearts that beat as one." There is nothing low or degrading about it, when it is what it ought to be, when it is brought to, and experienced at, its highest and best estate. It is *God-designed, God-born, God-bestowed!* As such it should be thankfully received and *divinely used* by all the sons and daughters of men.

VII
THE FIRST UNION

And now, although so much has been said, there is much that remains to be said, and which ought to be said, to do the subject justice. Some of these things are as follows:

Something more ought to be told about the second part of the act of coitus, the union of the organs, when this occurs for the *first* time on the part of the woman.

At the first meeting of the husband and wife, if the woman be a virgin, there are certain conditions which exist, on her part, that are not present in after-meetings, and these must be understood and rightly dealt with, or the worst of bad results may ensue.

Of course, at such first meeting, all the preliminaries prescribed as forming the *first* movement of the act should be carried out *to the limit*. It is not too much to say that these should be prolonged for *some days*! Do not start, young husband, at this statement! Well did Alexander Dumas, père, write: "Oh, young husband, have a care in the first overtures you make toward your bride! She may shrink from what she feels must come; she may put her hands over her eyes to shut out the sight; but do not forget that she is a woman, and so is filled with *curiosity*, under any and all circumstances! And you may set it down as sure, that, though she blinds herself with her hands as she scales the dizzy heights you are leading her over, nevertheless, *she will peek through her fingers!* So she will watch you with most critical eyes, and note every show of *selfishness or blundering on your part! So have a care!* You may think you are aiming your arrow at the sun. See to it that it does not alight in the mud!" Good words these, and to be heeded, come what may!

As a rule, if the bride be a virgin, it is well to *let plenty of time elapse before engaging in the full act of coitus!* Delay here will lead to a possible loving speed, later on. The young people should take time enough to get better acquainted with each other than ever before; to become, in a measure, accustomed to the uncovered presence of each other, and to the new possibilities of "courting" and "playing together" that their new conditions offer. In any case, full coitus should not be attempted till the bride is at least *willing*. If she can be brought to become *anxious* for the meeting, so much the better.

And so, with plenty of time taken for making ready for the act, we come to the first union of the organs for a newly married couple, the bride being a virgin. And here is where an explanation is called for.

The vulva, or external part of the female sex organs, is a mouth shaped aperture, located laterally between the forward part of the thighs. In shape, size and structure, it much resembles the external parts of the mouth proper. It begins just in front of the anus, and extends forward above the pubic bone and a little ways up the belly. Its entire lateral length is about four or more inches.

This organ is made up of several parts, as follows: The lips, or labiae, as they are technically known, the clitoris, and the vaginal opening. The lips are a double row, two on either side, and are known as labiae major and labiae minor, that is, the thicker and thinner, or larger and smaller lips. They extend almost the entire length of the vulva, the outer lips folding over the inner ones when the thighs are together. The outer parts of the larger lips are covered with hair. In thickness and quality these labiae are much like the lips of the face of each individual, a large mouth and thick lips indicate a large vulva and thick labiae and vice-versa. The clitoris is a gland that is located forward, on the upper part of the vulva. It corresponds, almost exactly, in make-up and function, with the glans penis of the male organ. The vaginal opening is at the rear, or lower part of the vulva, and leads directly into the vagina proper.

All these parts are composed of most keenly responsive nerves, and they are covered with a thin, delicate and exceedingly sensitive skin, almost exactly such as lines the cheeks and the mouth. Both the clitoris and the lips are filled with expandable blood vessels, and in a state of tumescence they are greatly enlarged by a flow of blood into the parts. The clitoris, in this condition, undergoes an enlargement, or "erection," which is exactly like that of the glans penis. So much as to the physiology of this part of the female sex organs, all of which should be well understood by every bride and bridegroom, though often it is not.

Now, in its virgin state, the vulva has another part, not yet named, and this is the hymen, or "maiden-head" as it is commonly known. This is a membrane that grows across the forward, or upper part of the vaginal opening, and so *closes up* nearly all that part of the vulva. This hymen is not always present, however, even in a state of undoubted virginity. Sometimes it is torn away in childhood by the little girl's fingers, as she "plays with herself." Sometimes it is ruptured by lifting, again it is broken away by the use of a large-sized female syringe. *For all these reasons, it is not right to conclude that a bride is not a virgin because the hymen is not present and in evidence at the first coition.*

Now many young husbands, and some young wives, are wholly ignorant of the *existence* of the hymen, and of the troubles it may cause at the second part of the sexual act, in a first meeting. This membrane is often quite tough and strong. It is grown fast to the lower part of the clitoris and to the inside surfaces of the smaller lips, and it covers so much of the vaginal opening that it is practically impossible for the erect penis to enter the vagina so long as it is present. Now if, under these conditions, the bride and groom (especially the latter) are ignorant of the real construction of the parts, and so should try to make a union of the organs, they would find such union obstructed, if not impossible; and if the man, puzzled, and impatient, and passion-driven, should *force* a hasty entrance into the vagina, rupturing the hymen ruthlessly, he would hurt the woman cruelly, probably cause her to *bleed* freely from the wounded parts, and shock her seriously! All of which would be a score against the husband, would brand him as a brute, or a bungler, and so tend to make his "sun-aimed arrow alight in the mud."

The thing to do here, is, first of all, to know the situation and to talk it over, and carefully, delicately, do the best that can be done about it. If the conditions are fully understood by the bride and groom, they can, in almost every case, by working and moving together carefully, overcome the obstacle, remove the hymen with little or no pain or loss of blood.

As a matter of fact, when the time for meeting comes, if all the facts are known, and the husband will hold his erect penis still and steady against the hymen, the bride will so press against it, and "wiggle around" it, that *by her own motions,* she will break the membrane and so be rid of it. She knows how much pain she can endure, and when the pressure is too hard she can relieve it by her own action! Anyhow, what is done *she does* herself, and so can never charge up against her husband!

It is a rare case in which, by mutual willingness, and desire and mutual effort to remove the obstruction, it cannot be eliminated with satisfaction to both bride and groom. If, however, careful and well-executed efforts fail to remove it, the services of a surgeon should be procured, and he, by a very simple and almost painless operation, can remove the difficulty. But never, *no never*, should it be brutally torn away by the force of the husband, and without the full willingness of the wife. *Mark this well.* As a matter of fact, the wise and practical thing for every bride to do, would be to go to a surgeon a few days before her wedding, and have him remove the hymen for her. Such operation is nearly painless, and is very easily done. Still, to do this might raise a doubt of virginity on the part of the husband and so this is a point to be careful about!

The act of removing the hymen is often spoken of as "defloration" — the tearing to pieces of a flower. The term is not fortunate. Nothing worth while has been taken away by removing the hymen, but much that is useful has been acquired. An organ that has outlived whatever usefulness it might once have had has been removed, and its going has made possible new and beautiful uses in life. If this has been accomplished by the mutual desire and effort of the bride and groom, it is a cause for joy and not of sorrow; of delight and not of mourning. As well weep over the removal of the vermiform appendix as for the destruction of the hymen.

With this obstacle rightly overcome, the second act of coitus offers no situation that calls for further remark or explanation.

And now a few words about the probabilities of conception resulting from coitus, and some matters which are very closely related thereto.

In the first place, every healthy and fairly-well-provided-for husband and wife should desire to have children, and should act in accordance with such wish. This is not only in harmony with the primary purpose of sex in the human family, but it is a response to a natural demand of the human soul, in both man and woman. As Bernard Shaw makes Jack Tanner say: "There is a father-heart as well as a mother-heart" and *parenthood is the supreme desire of all normal and wholesome-minded men and women.* It is not an "instinct," but something far above that quality.

Parenthood among mere animals is the result of instinct, and of that alone, but not so in the human race. Human beings naturally desire to make a home for themselves, and a home, in the fullest meaning of that word, means *children* and a "family circle." This is something that animals know nothing about. Animal mothers forget and ignore their progeny as soon as they are weaned; and animal fathers will, in many cases, kill them as soon as they are born, if they get a chance to do so. These facts prove that parenthood, in the human family, is something much more than in the rest of the animal kingdom. Indeed, the whole matter of comparing this quality, as it exists in humanity, with that of animals merely, is only a continuance of the similar abomination of comparing the sex functions of these two forms of life. In the real essentials of existence, they are in no way comparable; and to make such is not only folly, but approaches the positively criminal. The results of doing so certainly lead to crime.

Fundamentally, then, nearly all men and women marry with the purpose and hope of having a family of children. They may not put it that way, may not even acknowledge it, even to each other or to themselves; but if married people find that they *cannot* produce, it is a source of unspeakable regret to them both. In such cases, the inherent desire for parenthood will "cry aloud and spare not." A "barren" woman greatly mourns her inability, and will shed bitter tears over the fact, if she be truly human; and an "impotent" man will be practically despised by all who are aware of his incompetence.

And yet, though all normal men and women desire to have children, it is only right that they should desire to have them *as they want them*, and *when* they want them, and not *whenever they may happen to come!* That is, sensible and thoughtful people, who plan definitely for the future, want to make the coming of children to them an affair of *deliberate* arrangement, and not of *chance*.

This is not only as it should be, but is really the only right way that children should be begotten and born. Which statement calls for a few special words on the right of parents to regulate the production of progeny.

There is much talk, in some quarters, about "race suicide," and the wickedness of deliberately limiting the number of children in a family. Such talking and writing arouse anxious questionings in the minds of conscientious young married men and women who are desiring to do the right thing in the premises, but are uncertain as to what the right thing is, and for such are the following words:

Many years ago, an English philosopher and statesman, Malthus by name, discovered and announced the fact that the rate of natural increase in the human race was several times greater than that of the possible rate of production of food supply for their support. Scientifically phrased, his statement was that "the rate of increase in humanity is in geometrical ratio, while the rate of increase of possible food supply is in arithmetical ratio." And from this basis, he reasoned that, unless the surplus of human production was in some way cut off and destroyed, the whole human race would ultimately demand more food supply than could possibly be produced; and so, in due course of time, the whole race would perish from starvation!

Then he proceeded to reason that the purpose of disease, plague, pestilence, famine, poverty and warfare was to cut off and destroy the *surplus* of humanity, and hence all these alleged evils were in reality blessings in disguise, and that *it would be wrong to interfere* with their really beneficent workings! Volumes could be written, and they could not tell the half of the misery and evil that the promulgation of this doctrine has done for the civilized world, but there is no space here for giving any such details; nor need this be done, though the statement of the doctrine had to be made to make ready for what is to follow.

Now, is it not far more reasonable to suppose that, *since the possibility of determining the number of offspring a husband and wife may produce has been given them;* that since such result can be, for them, made a matter of *choice,* of an *exercise of the will,* and not of *blind instinct* — under these circumstances, all of which undoubtedly exist, is it not far more reasonable to believe that it is the *purpose of the Creator* that the limiting of the number of human beings in the world should be brought about by *curbing the birth rate,* rather than by *killing the surplus* after they are born!

There can be but one answer made to this question, by any intelligent man or woman.

These facts, then, establish the *rightfulness of determining the number and size of a family by every husband and wife.* But this does not mean that they are to entirely refrain from cohabiting, in order to keep from having children! This phase of the argument has already been gone over and disposed of. But it *does* mean that husbands and wives have a right to use such rightful means for the limiting of the number of offspring as are conducive to the interests of all parties concerned — themselves, their circumstances, the born or unborn children, the state, the nation. Let the bride and groom be well convinced and established in their own minds on these points, as early in their relation as possible. They should be so from the very outset — *must* be so, to reach the best results.

The issue then presents itself: How can such deliberate and wilful determination of the number of children a husband and wife may have, be brought about?

And the answer is, that *it can never be accomplished by careless and hap-hazard cohabiting!* On the contrary, it can only be compassed by the most *careful* and *watchful* processes of engaging in coitus, and by a *full knowledge* of physiological facts, and by acting, *always,* in accordance with the same. It is no road for careless travel, but it is a way worth going in, for all that.

On this point, let it be said that all sane and intelligent men and women agree that anything even approaching *infanticide* is nothing short of a crime, and that abortion, except for the purpose of saving the life of the mother, is practically murder.

But, while this is all true, to prevent the contact of two germs which, if permitted to unite, would be liable to result in a living human form, is *quite another affair*.

It is only this aspect of the situation which will be considered in what follows.

Now, as has already been shown, the essentials for conception consist of having the ovum present in the womb, and its meeting the semen there. The corollary of this is, that whenever these coincidences take place, there is a *possibility* for conception.

But in all *normal* cases, the ovum only passes into the womb once in every twenty-eight days; and, as a rule, it only remains in the womb for about half that period of time, that is, for about 14 or 15 days in each month. And so, since the menstrual flow ceases after about five days from its beginning, in about ten days *after* its stopping, the ovum will have passed out of the womb, and hence that organ contains nothing that is impregnable. Under these conditions, semen may be deposited in the womb, without danger of impregnation. This is a simple proposition, and easy to understand if once known.

However, it must be said that these *generally* common conditions *do not always obtain* — that is, they are *not* true in the case of *all* women. There are women who will conceive at *any* time in the month, if they are given a chance to do so. The physiological reason for such possibility is said to be this: There are always ova in the ovaries, in varying stages of development. Ordinarily, only once a month do any of these pass down into the womb; but, in exceptional cases, sometimes these ova are so partially held in the ovaries that, under the excitement of coitus, and because all these parts dilate so much during the act, an ovum may slip its moorings, under such conditions, pass down into the uterus at an untimely season, meet the semen there, and pregnancy result. Such are the facts *in some cases*.

How, then, can a husband and wife tell how it is, or will be, in *their* particular case?

The answer is that they can only tell by trying, and that should be done as follows:

The *first* sexual meeting of the bride and groom should *never* take place until at least *ten days after the ceasing of the menstrual flow in the bride! This is a rule that should never be violated* if the parties wish to "*test out*" the real condition as to whether or not the bride has any "free time." The chances are several to one that she *has* such leeway; but the fact can only be established by "proving up" and this can *never* be done if any *chances* are taken. Put this down as rule number one.

For this reason, it is well for the bride to fix the wedding day; and, if possible, for her to locate it sometime during the probably immune period. And the nearer she can bring this day to the *beginning* of such period of freedom from danger of pregnancy, the better. For, if it should happen that the first coitus should take place only a *day or two before* the time when another "monthly" was due, such excitement might hasten the passage of the nearly-ripe ovum into the uterus, and conception might occur. In which case, "all the fat would be in the fire," nothing would be proved, and the parties would be as ignorant as ever regarding the facts in *their* case.

And so, the *first* sexual meeting of a bride and bridegroom should be not *earlier* than *ten days after the ceasing of the menstrual flow and not later than three days before the next monthly is due. Put that down as rule number two, never to be violated.*

And if marriage takes place before this period of probable immunity on the part of the bride arrives, the only safe thing to do is to "patiently wait" till such time arrives. This may "require fortitude" on the part of both parties, but it is the only safe thing to do. And to do just that, will amply repay such waiting. The writer knows of a case where the wedding took place just three days before the bride's next monthly was due, and she and her husband waited for more than *two weeks* before they met sexually! But it paid to wait, for their doing so proved that the bride had *two weeks* of "*free time*" in *each month, and this was worth all it cost to find out! Take time!*

And now let it be added that it is a great accomplishment for a husband and wife to be free from a fear of pregnancy as a result of coitus. This is a thousand times truer for the woman than for the man, for it is she who has to bear the burden of what follows, if following there be. The husband can "do the deed" and go about his business. The wife, if "the fertile seed" takes root, has before her months of care and anxiety, and she risks her very life in what may come of it all. For these reasons, she has a *right to dictate all the terms* which are liable to cause her to become a mother. *And yet she should do this with full regard for the husband, in love, in true wifely-womanhood.* On this point, do not fail to read "The Helpmate," by May Sinclair. It is a story that no bride and bridegroom should fail to read and study, carefully.

The whole subject of how to engage in satisfactory coitus and avoid pregnancy may be summed up as follows: — The attainment of such a condition is well worth the most careful, earnest and honestly pains-taking endeavor. For, if such status be not reached, its lack will be a source of endless contentions and differences between the husband and wife. It will lead to jealousies, quarrels, and all sorts of marital woes. But, the situation once mastered, by the most loving and accurate of scientific methods of procedure, a happy married life is certain to result. Otherwise, the "married state" will always be in a condition of "unstable equilibrium." So let every bride and bridegroom begin, *from the first*, to try to establish the greatly to be desired accomplishment. If anything further on this point should be desired, consult a reliable physician.

VIII
The Art Of Love

And still there is more to be said! Is it not written that "Art is long!" *And the Art of Love is the longest of all arts, and the most difficult of all for its complete mastery and attainment!*

It is a matter of misfortune, and yet one of not infrequent occurrence, that the sex organs of husband and wife are *not well matched*; and that trouble, sometimes of a most serious nature, results. When this condition is found to exist, it should be treated sanely and wisely, and the chances are many to one that the difficulty can be overcome, to the full satisfaction of both parties concerned.

In such cases, the mis-matching usually arises from the fact that the penis of the husband is too long for the vagina of the wife. This is very apt to be the case where the wife is of the "dumpy" sort, with a small mouth and short fingers, while the husband is "gangling," large mouthed and long fingered. These are facts that ought to be taken into account before marriage, and which should figure in determining whether the parties are "suited" to each other. They *would* be regarded in this way, too, if they were generally known, as they most surely are not. Here is another place where ignorance and "innocence" get in their work, and make trouble in married life!

In such a case as this, the too-long penis, when fully inserted in the too-short vagina, and especially when, at the orgasm, the two organs are crowded together vigorously, as the impulse of both parties demands they should be at this part of the act, the end of the penis is driven against the rear walls of the vagina, often furiously, thus stretching and straining the vaginal passage longitudinally, pressing against the womb unnaturally, and not infrequently pushing it out of place and sometimes rupturing the uterine tract seriously, hence causing all sorts of unfortunate and greatly-to-be-regretted results.

Because of such danger, the first meeting of the husband and wife should be accomplished with the utmost care, especially in the *second* part of the act, the first putting together of the organs. This is the only way of determining, in each case, how the organs will "fit," and happy are the parties thereto if such fit is found to be perfect!

But if it should turn out that there is a mismatching, of the nature just described, the conditions can be adjusted if the right means are used.

(Before telling this, however, it should be stated that the relative size of the sex organs can never be fully judged of by the size of the body of a man or a woman. Many a small man has an abnormally large and long penis, and many a little woman has a large vulva and a long vagina; and the reverse of all this is true, in the case of many men and women. These items in the count are among the things that can never be known with certainty except by actual trial, and this is not possible, as things are now.)

And so, if "mis-matching" is found to exist, in any given case, it can be provided for, in most cases as follows:

Instead of taking the position for coitus which has already been described — the woman on her back and the man over and above her — let *this* be done: Let the man lie on his left side, or partly on his left side and partly on his back, facing the woman, his left leg drawn up so that the thigh makes an angle of 45 degrees with the body, and the knee bent at about the same angle. Now let her, lying on her right side, mount into his arms, in this way: Let her place her right hip in the angle made by her husband's left thigh and his body, so that *his left leg* supports *her hips,* by being under them; put her right leg between his legs, throw her left leg over his right leg, put her right arm around his neck, and her left arm should be placed across his body under his right arm. His left arm should be placed around her waist from below, and his right arm left free to move over her body, as he may choose. Now in *this* position, the man's hips make a sort of saddle into which the woman "vaults" easily, naturally, and with the greatest of comfort; while the man, with his whole body supported by the bed, as he lies, will be perfectly comfortable, and can maintain the position much longer, without tiring, than he could were he over and above the woman, supporting himself by his elbows and knees, and with the woman's arms around his waist, lifting her body thereby, and thus adding her weight to his, all to be sustained by him. A moment's consideration will disclose the fact that this position has many points in its favor, beyond that of the man-superior form. The woman, in this position, is not wholly superior, but she is partly on her right side and partly on her belly. Her whole weight rests on her husband's body, but her weight does not tire him, as the bed below him easily supports them both.

Now, in this position, the sex organs are brought closely together and their union is easily accomplished. But see! It is *now* the *woman,* and not the *man* who has *full control* of such meeting, and so can regulate it to *her liking,* or *needs.* Her hips are perfectly free to move towards, or from, those of the man; and so *she can determine just how much or how little of his penis shall enter her vagina!* And if his penis is too long for her, she can accommodate her action to such fact!

As for the man, his satisfaction will be fully equal to, if not greater than it would be were he in the other position. The ease afforded to his body, and the fact that he need have no fear of hurting the woman, these things will be a delight to him, that is of real value, and which will make for his delectation as much as for that of the woman in his arms. The in-and-out motion is as easily performed in this position as in the other; and at the climax, the organs can be crowded together passionately, and still without hurting the woman. For she, being free to move, can so curve her hips that the pelvic bone, the *mons veneris,* as it is technically called, will receive the most of the pressure, and at the same time the angle which is thus made by the relative positions of the vagina and the penis will keep the latter from penetrating the vagina too far, and so will protect its rear walls and the womb from all danger of harm. The orgasm is just as perfect in this position as in the other. It is just as *natural* as the other position, and has only to be tried to be proved worthy.

And now one other point. (Curious how these details protract themselves. But there is no help for it. We must continue, now that we have begun.)

A very frequent cause of married unsatisfaction is the fact of the *difference of time* that it takes for the husband and wife to come to the climax, the orgasm. As has already been noted, the highest delight in the act comes when this climax is simultaneous, comes at exactly the same instant to both parties. But to bring this about is not easy in all cases, and hence what follows:

As a rule, women are slower in reaching the orgasm than are men. This is not always so, but it is generally the case. Some wives are so passionate that they will "spend" several times to their husbands' once! The author knows of a case where the wife will regularly experience the orgasm four or five times to her husband's once. She is a lovely wife and a highly accomplished woman, in no sense "fleshy" or "worldly minded." The situation is that her sex organs are exceedingly sensitive while those of her husband are the reverse, they are "timed" differently, that is all. The case is rare, and as a rule, women are "timed" slower than men.

Again, after a man has passed the orgasm it is, in most cases, impossible for him to continue the act, right then and there, and bring the woman to the climax, if she has not yet arrived, from the fact that, with the expulsion of the semen, usually detumescence of the penis at once takes place, and the organ is incapable of exciting the woman when in this condition. And so, if the husband "goes off" *first,* there is no possibility of the wife's reaching the climax at that embrace. This leaves her unsatisfied, all her sex organs congested, and the whole situation is unsatisfactory, in the extreme. On the other hand, if the wife comes to the orgasm first, her vulva and vagina detumesce but little and that very slowly, so that it is perfectly possible for the husband to continue his action, and come to the climax, even if his partner has already "spent."

Under these conditions it is easy to see that, where the wife is "keyed" or "timed" much slower than her husband, as is quite often the case, coitus is very liable to be a very one-sided affair, one in which the *husband gets all the satisfaction, and the wife little or* NONE — *a most unfortunate status for both parties, but especially for the wife.* The writer once knew a case where a husband and wife lived together to celebrate their golden wedding, and the wife never once experienced an orgasm, though the husband cohabited with her several times a month, during the most of their married life. There was no good reason why this should have been so, only that the husband was "quick in action" and the wife somewhat slow, and they had never synchronated their time differences. The dear old lady died at ninety, never having known a joy that, since her bridal night, she had wished for. Both the husband and wife were most excellent people. *They simply didn't know!* One was ignorant and the other innocent, and there you are again!

Now the thing to do, under such circumstances, is for the parties to "get together." And the way to do this is, first, to *prolong the FIRST part* of the act, till the wife has not only caught up with, but is even *ahead* of her husband in the state of her passion. To bring about this condition, *the husband should use every means to stimulate his wife's sex-nature and increase her desire for coition.* Here are some things he can do, which will tend to produce such results:

A woman's breasts are directly connected with all her reproductive nerves. This is especially true of her nipples. To touch them is to directly excite all of her sex organs. The lips and tongue are also thus nervously connected with these vital parts, and, so, if the husband will "play" with his wife's breasts, especially with her nipples, manipulating them with his fingers, or, better still, with his lips and tongue — at the same time, if he will stroke her vulva with his fingers, especially the clitoris, *and if she will encourage him to do this,* by holding her breast with one hand, shaking it about as her nipple is in her lover's lips; if, lying flat on her back, her husband at her right side, and with his left arm around her waist, she will spread her legs wide apart, thus opening the vulva to its utmost, and sway her hips, raising and lowering them betimes; and, since she has a free hand, if, with this, she will take her husband's penis with it and "play" with it as her lover plays with her vulva – if they will do this, the cases are rare in which passion will not grow in the wife to almost any desirable extent. Under such "courting," the parts will all enlarge, the pre-coital secretion will flow in abundance; and, in due course, all will be ready for the second part of the act. This part of coitus is, really, one of the most enjoyable of the entire performance.

If, perchance, the pre-coital secretion should be tardy in appearing on the part of the wife, so that the vulva is dry as the husband strokes it, let him moisten the part with saliva from his mouth. To do this, let him moisten his *fingers* from his mouth, and transfer this to the vulva, and then proceed with his stroking. This moistening the vulva with saliva may be repeated *several* times, *if necessary,* always until the flow of pre-coital fluid from the parts themselves renders any further moistening needless. *The stroking of the dry vulva will do little toward the arousing of passion, or producing the pre-coital flow.* But if the parts be moistened, as above directed, both these desired results will follow, except in *very* rare cases.

And let no one make the mistake of thinking that thus moistening the vulva with saliva is unseemly, or unsanitary. It is neither. On the contrary, it is nature's way of helping to perfection an act which, but for such timely assistance, might never be brought to a successful issue. As has already been noted, chemically, saliva and the pre-coital fluid are almost identical. They are both a natural secretion of a mucous membrane, are alkaline in reaction, their native purpose is lubrication, and, as a matter of fact, the saliva is as natural an application to the lips of the vulva as it is to the interior of the mouth or throat. Truth to tell, the practice of applying saliva to the genitals before coition is very general, so much so that it might almost be counted as instinctive. It is mentioned here only to remove any prejudice that might linger in the sophisticated mind of the reader. Such use of saliva is no more to be deprecated than its application in a hundred other ways, such as moistening the fingers to turn a leaf, of "licking" one's fingers after eating candy. Such use of this fluid from the mouth might be condemned by the "over-nice," but it is quite universally practiced, and it is neither unwholesome nor unsanitary.

It is sometimes recommended that some form of oil, as sweet oil or vaseline, be used as an unguent for anointing the parts before engaging in coitus, but this practice cannot be recommended. Oil is not a natural product of the parts to which it is applied, it is chemically unlike their secretions, and to smear the delicate organs with a fluid that is foreign to their nature, is unwise, unsanitary, not to say filthy. It is like greasing the mouth to make food slip down easily. And it is easy to understand how such application of an unguent to the mouth would impair the taste, dull the nerves of sensation, and greatly interfere with the native and wholesome uses of the oral cavity.

So don't be afraid or ashamed to use saliva in preparing the vulva and the vagina for the reception of their natural mate.

And so, to return to where we left off, if the wife is slower timed than her husband, her passion can be greatly increased by the manipulation just described. Indeed, it could be very easily carried to such length — the lips and tongue playing with the nipple, and the finger-stroking of the vulva — that the woman could be brought to an orgasm without the union of the organs at all! This is a form of masturbation (this word has a bad meaning attached to it, but it is a good word, as will shortly be shown, and it has its legitimate uses; but, as a preparation for coition, it should not be carried any further than is essential for bringing the laggard passion of the woman up to an equal tension of that of her lover.) A few weeks', or months', practice will enable a wife to determine just how much of this form of "courting" will bring her to the desired point of excitement; and, when this point is reached, she should invite her husband to "come up over," if the first position is to be adopted for the rest of the act; or, she should throw herself into her lover's arms, if the second position is used.

Just a little more — If, after getting into one position or the other, it seems to the wife that she is not yet fairly abreast of her husband in the intensity of her passion, let her *still further* seek to advance it, as follows:

If the position with the husband superior is taken, let him, after he has gotten into place and before the organs are united, have his wife take his penis in her hand, and, as he moves his hips up and down, stroke her vulva, especially the clitoris, with the glans penis — not entering the vagina at once, but continuing this form of *exterior* contact of the organs, for a longer or shorter time — slipping past the wide open vaginal mouth, even when the wife raises her thighs and, as it were, begs for an entrance; tantalizing her to the point of distraction — -till, finally, she will "take no for an answer" no longer, but will, in an ecstacy, slip the penis into the vagina, and thus consummate their union.

If she be far enough abandoned with her passion, such entrance may be made at a single stroke, not to say a furious plunge. But if the vulva and vagina are not yet fully dilated, the entrance should be carefully made, gently made, as she can bear it, as *she* wishes it to be.

Sometimes, yes, not infrequently, in this position, the external stroking of the organs may be continued to the very verge of the orgasm, so that, especially if the entrance can be made, as it were, in a frenzy of passionate delight, the organs coming into full length union at a single impulse, or rushing together — then the simultaneous climax *may* be reached with one or two in-and-out motions — or, perhaps the single master-plunge may win the goal instanter! If so, a consummation devoutly to be wished has been successfully reached!

Again, if the wife is slow, and the man is quick, in this play for "getting together," it will enable the man to greatly extend and protract what might be called the time of his possible *retention*, if he can keep the foreskin over the glans penis. Some men cannot do this. If they have been circumcised, of course they cannot! But if the glans penis can be covered with the foreskin during all this playing together, it will enable the husband to prolong his "retentional time" far beyond what he otherwise could. Some men have the power of "retaining" to almost any length of time by the exercise of their will power, and so they can *wait* for their wives. If the wife is slower timed than the husband, he should *carefully cultivate the "art of retaining"* and so wait for her. *To do this successfully will greatly increase married happiness.*

This same remark (keeping the gland covered) applies with equal force to the possibilities of the man's retention after the organs are united, and all through the third part of the act. If the penis can enter the vagina with its "natural cap on," the husband can give his wife the pleasure of many times the amount of in-and-out motion than he could otherwise bestow upon her. And if the wife is the slower of the two (as is generally the case) she will greatly appreciate such a favor, and will repay it a THOUSAND FOLD by the responsive, reciprocal motions which she will LAVISH upon her *considerate* lover.

This is an item of almost supreme importance — this "keeping the cap on" the penis, during the act, *if the wife is slower than the husband* — if they need to have a care, to insure their "getting off together."

And here is a curious fact, which would seem to show that Mother Nature has especially provided a blissful reward for both the husband and wife who will be careful on this point. Thus, if the husband will be careful to have the glans penis covered with the foreskin (and, of course, this can *never* be, if the organs are united when the vulva and vagina are dry) when it enters the vagina, and will so engage in the in-and-out motion that it will *stay covered* as the *third* act progresses — if this is done, when the climax comes, if the two "spend together," the womb will open its mouth as it were, clasp the foreskin, slip it back over the gland so that, when the supreme instant comes, the naked gland will be in the most direct and blissful contact with the most sensitive part of the uterus! This is a most wonderful provision of nature, and to utilize it, and enjoy it to its utmost, is the maximum of human delight!

Again, if after the organs are well together, in the man-superior position, and the in-and-out motion has begun, it should be found that the wife is still behind in the game, she can gain greatly in "catching up" if she is permitted to *originate* the larger part of the motion. To enable her to do this, let her husband hold his body quite well above her, so that she can have plenty of freedom to move her hips as she may choose to. Added to this, if the husband will, in large measure, "hold still," and keep his penis in such position that it presses against the *upper part* of the vulva, that is against the clitoris, (as the phrase goes, if he will "ride high") and then permit his *wife* to make "long strokes," sliding the organs together for their full possible length, with the clitoris in constant contact with the penis, during the whole of each stroke — all of this will greatly and rapidly increase her passions and bring her to the climax.

Or, as a variation from this, if the organs can be united to their fullest possible limit, so that the base of the penis presses firmly against the Mons Veneris, and the clitoris and labiae almost clasp their mate; and then, in this position, if the husband will maintain the *status quo*, while she lifts her hips hard against his, and *swings them about*, in a sort of circular motion "round and round," as it were — this will also greatly increase her passion, and soon bring her to the climax.

In both these last described ways of courting, the husband should be *extra careful not* to permit the weight of his body to press down heavily upon his wife. He should *wholly* sustain himself on his elbows and knees, and permit her to lift herself, at least her hips, by the help of her arms around his waist. This is no hardship for the husband, if he be a true lover. For is he not strong, and what is his strength for but to delight his sweetheart? *A true, devoted, virile and manly lover is always at the service of his sweetheart! To delight her, is to doubly delight himself.* This is another point of which mere animals know nothing. There is nothing in all their nature which responds to the like of this, in any way. The whole experience is *human;* it is productive of a joy, of a *spiritual elevation,* which mere animality knows nothing of — can know nothing of.

Playing thus together, courting each other thus (For, through all these actions, a line of *complete mutualness must run*! The husband may *seem* to be specially accommodating himself, and all he does, to his wife's whims or necessities; but, even so, this will be more of a delight to *him* than it is to *her,* viewed from the *spiritual plane,* on the principle that "it is more blessed to give than to receive" — and no truer words than these were ever spoken — while, at the same time, the wife, though *seeming* only to be gratifying herself, to be reaching after what she alone desires, yet, as a matter of fact, by her very so doing — and the more perfectly, completely, she does this, the better — she is gratifying and delighting her husband to the utmost possible limit) courting each other thus, the lovers will learn to "time" themselves together, perfectly, each knowing just when the other is fully ready, by a sort of *spiritual consciousness,* as it were, and so a perfect climax can be reached.

Take time, LET LOVE RULE AND DIRECT; BANISH ALL SELFISHNESS; *Let the husband keep his head, and* THE WIFE UTTERLY LOSE HERS, throwing it to the winds, to be wholly swept away by the whirlwind of her passion; feeling free, delighting, to let it go, go, go, no one cares where! Do these things, and married life will be glorious! Of such is the kingdom of heaven, for the truly wedded lovers!

This will be "all Greek," or "foolishness" to the selfish and materially-minded; but to the truly wise, it will be *life immeasurable*. This is a paradox, but it takes a paradox to tell the greatest truths!

So much for the act of coitus in the man-superior position, when the wife is slower timed than the husband and they adopt this method, and the accompanying means for "getting together." Now, if the other position is taken, that of the wife semi-superior, in the husband's arms, as he lies partly on his back and partly on his left side, etc., here are a few points to be noted to advantage.

Still assuming that the wife is the slower-timed of the two, it is entirely possible that when she has "come over" and has gotten into position, that she may not yet be fully ready for the union of the organs. The very time that it takes for her to get into position, the changing of the position of her body, from her back to her right side; the temporary cessation of the stroking of the vulva by her husbands's [sic] fingers; all these things will have a tendency to retard her passion, for the time being, and all this loss ought to be made good, if not added to, before the *second* part of the act is entered upon. And, in this position, all this can most happily be brought about, as follows: —

Lying in each other's arms, in this *second* described position, the organs naturally *come* into contact in such a way as to make the further excitation of the vulva and clitoris most natural and easy. The spreading of the wife's hips, caused by her throwing her left leg over her husband's right and drawing up of her left knee, opens the vulva wide; and, at the same time, the penis, from the very nature of its position, will lie at full length in the opening, thus exposed — not entering the vagina, but remaining "without the gate" as yet.

By this time the vulva will have become enlarged and elongated, the lips full and the clitoris erect, all in a state of tumescence, and all covered with the pre-coital fluid; the lips so distended that, when thus parted, they form the sides of a labial canal, as it were (a delectable, and most delicately smooth-walled channel). Now, in this extended condition, which is fully as long as the penis, from end to end of its pathway of dalliance, every part covered with the most delicately sensitive nerve-filaments, and all of these in an ecstasy of keenness to the sense of touch, and in the most perfect of "love's strolling way," — if the penis, as it were, stands up full and strong, in such fashion that it touches the vulva at every point, both inner and outer labiae, the clitoris and all, for a space of five or six inches in length; while the protruded and well-moistened lips of the vulva as it were reach out, and clasp themselves at least half way around their suitor, laving him with their luscious kisses — in this position, the wife being partly above, and so, perfectly free to move her "love way" as she will, she can slide the pathway itself a full six or more inches, up and down, stroking all the area against the penis as she moves; that, again, by its very position, being held firmly in contact by its stiffness and stoutness; the glans penis throbbing lustily against the clitoris when the two meet at the extreme of the wife's up-stroke; she, pausing an instant, just then, to more perfectly enjoy the sensation; the penis slipping past the now wide open vaginal mouth, which reaches out at every down stroke to engulf it — dallying, delaying, coquetting, tantalizing, both man and woman; playing the game in almost a swoon of ecstatic delight — under such conditions the wife's passion will rush to its fullest development, till, when she will, she can drop her vagina upon the penis in such a way that the *two will be made one,* in absolute perfection, on a single move, and from this to the finish it is but a few motions distant.

In some respects this manner of coitus, and this means of "going off together" is unsurpassed.

Which leads to the remark that this position is sometimes the best for the full completion of the act. It is the easiest of all positions, the least fatiguing. And if the wife is tired, or not quite "up to grade," she can enjoy an embrace of this sort without fatigue, even to the full. For the organs can be united in this position quite perfectly, though the penis will not penetrate the vagina to as great a length as in the other position. Still, the climax can be perfectly reached in this way, and it is one of the best ways to make sure of perfect "timing," of "spending" exactly together, which is greatly in its favor.

If there is a mis-matching of the organs, the vagina of the wife being too short for her husband's penis, this is a most excellent way for meeting and overcoming that difficulty.

This naturally leads to another matter, as follows: — It might seem to the reader that the different "strokings" of the vulva, with the fingers, or the penis, all the contact being outside the vagina, that all of these methods of excitation smack of masturbation, and so are of doubtful rightness. In reply to which, note the following:

The entire affair of coition, in humanity, has already been shown to be something wholly above and beyond mere animality. It is the exercise of functions that belong *only to mankind,* and hence is not amenable to *any* merely *animal* laws or restrictions! It is the source of numberless human joys, and *any* method of engaging in the act of mutual delight, that is, of *mutually happifying*, is legitimate and *altogether right*. And so, if the parties choose to increase their mutual delight, if the husband wishes to arouse and intensify his wife's passion by stroking her vulva with his saliva-moistened fingers, and *she wishes him to do so*, such act is as right and as wholesome as is coitus in the by-some-supposed-to-be *only* way of its exercise. Let this never be doubted.

The fact is, this whole matter of sexual excitation by means of the hand, or in other ways than the union of the organs, has received a black eye at the hands of would be purists, which it in no way deserves. As already noted, the word masturbation has been fastened to such acts, and then, any and every form of it has been condemned far beyond what the facts warrant, till the minds of the rank and file are wholly misled in the premises! When one looks at the situation from the point of view which insists that *all* the sex functions should be under the control of the *will*, then light is thrown upon the entire subject. Seen in this way, *any* form of sex stimulation, or auto-erotism even (auto-erotism means *self* sex-excitation) which is NOT CARRIED TO EXCESS, is *right* and *wholesome*! But we have been taught the contrary of this for so long that it is difficult for us to realize that it is true. *But it is*!

Hence, if it should sometimes happen that the husband should arrive at the climax before the wife does, and he could not bring her to an orgasm by excitation with his spent penis, it would be *perfectly right for him to substitute his fingers, and satisfy her in that way*. Of course, this would not be as satisfying to her as it would have been could she have met him simultaneously, but it is *far better than for her not to be entirely gratified! Many a woman* SUFFERS ALL NIGHT LONG *with unsatisfied desire, her organs congested and tumescent, because she has been left* UNSATISFIED *by a husband who has spent before she was ready,* AND THEN LEFT HER! Such cases might be *entirely relieved,* if the parties *knew the truth,* and were not too *ignorant,* or *prejudiced,* or *ashamed* to do what should be done to make the best of a situation.

Of course, no husband should make a *practice* of gratifying himself fully, and then bringing his wife to the climax with his fingers. Such a practice would be *selfish* and *wrong*. But as an *emergency* way of escape, the method is to be commended.

Of course, as has already been explained, the husband always has the advantage, that he can be brought to the orgasm by the insertion of the penis into the vagina, *after* his wife has spent, if she arrives first, since her organs detumesce slowly, and their distended condition permits such action on his part, for some time after she has passed the climax. But not so with the husband. Once spent, his penis shrinks to limpness, almost immediately, and in this condition it cannot satisfy the wife in the least, much less bring her to an orgasm.

Again, if, for any reason, the wife should be unable to meet her husband in coitus proper, because of weakness, or slight illness, or perhaps some temporary soreness of the parts, it would help the situation wonderfully if *she* would take *his* penis in *her* hand and "play with it" till he *spent*. He would love her for it, kiss her for it, give her his soul for it!

If a bride and bridegroom knew enough to introduce each other to the delights of an orgasm by "spending" each other by external excitation of the organs with their hands a few times before they united the organs at all, it would be to their lasting well being. This is especially true for the bride. If her lover would take her in his arms, even with all her clothes on, as she sat on his lap, in their bridal chamber, alone, and stroke her vulva till she "*spent,*" the chances are many to one that he would have introduced her to such a joy that she would never forget it, all her life. Surely, such method is *infinitely superior* to *raping* a bride, as is so frequently done by the ignorant or goody-good young husband, who "stands upon his *rights*!"

Indeed, if a bride to be, who was so innocent or ignorant of her own sex possibilities that she had never experienced an orgasm – had never "spent" – could be "put wise" before her bridal-night, if she could be instructed enough to lead her to engage in some form of auto-erotism, bringing herself to an orgasm with her own hand, *just for the sake of the experience it would give her, and so that she would have some clear idea of what she really wanted, before she went into the arms of her lover – if she could do this, in the right mental attitude, it would be greatly to her well-being, a worthy and valuable addition to her stock of knowledge of herself and of the powers that are latent within her. Her alleged loss of innocence by such act would be as nothing compared with the wisdom she would gain by the experience. When innocence leads to harmful results, it is time it was ended, and that knowledge takes its place!*

As for the husband, the chances are not one in a million that he will be ignorant of what an orgasm is like before he marries, since all healthy young men "spend" at least once a week, automatically, if not otherwise!

Let it be said further, that auto-erotism, self-spending, may be practiced by both men and women, to their healthful benefit, when sexual exercise cannot be secured in any other way. It is only when *carried to excess* that such action is in any way harmful. The only danger is, that, the individual being alone and having all the means for self-gratification in his or her own hands, so to speak, it is quite possible to indulge in the action too freely, which, of course, leads to bad results. *But the act itself is not bad.* On the contrary, when kept within bounds, it is healthful and wholesome.

There are many unmarried women, maiden ladies, and especially widows, who would greatly improve their health if they practiced some form of auto-erotism, occasionally. When husbands and wives are forced to be much away from each other, it is right for them to occasionally satisfy themselves in this way, their souls filled with loving thoughts of the absent one the while.

There is any amount of nonsense current about auto-erotism. As a matter of fact, all boys masturbate, and many girls also. Some authors claim that more than half of all women engage in some form of auto-erotism, at some time in their lives, and the estimate is probably too low rather than too high. But, unless they carry the act to excess, they are guilty of no wrong. Not infrequently, they may make the act a means of great good to themselves. *The sex organs are alive! They constantly secrete fluids that need to be excreted, as all other organs of the body do. They ought to be relieved, as their nature requires they should be.* If this cannot be accomplished as the most natural way prescribes, it is only right to do the next best thing. Only, it should not be carried to excess. Be temperate in all things. Gratify yourself, but don't ABUSE yourself.

Auto-erotism, or masturbation, should never be permitted to become "self-abuse," nor is there any need that it should ever do so. It should be self-upbuilding, not self degrading. Rightly used it can be thus.

IX
Coitus Reservatus

This brings us to another item in the matter of sexual exercise on the part of the husband and wife, as follows: –

It should be the constant aim and endeavor of both parties to continually lift all sex affairs above the plane of animality, mere physical gratification, into the realm of *mental* and *spiritual* delight. To this end, let it be said at once that such a condition can be reached, in the greatest degree, by the practice of what is known, in scientific terms, as "*coitus reservatus,*" which, translated, means going only *part* of the way in the act, and not carrying it to its climax, the orgasm. Described in terms with which the reader is now familiar, it means, carrying the act only through the first and second stages, the "courting" stage, and the union of the organs, and stopping there! This may seem, at first thought, neither right nor wise, but, as a matter of fact, it is both, as thousands of most happily married people have proved.

Going a bit into details, this act of "reservatus" really unites the first two parts of the act into a common whole, making it simply one continuous piece of "courting," merely that, and nothing more. It is almost entirely a *mental and spiritual love-embrace; and in its perfection, it exalts the husband and wife to the topmost heights of mental and spiritual enjoyment and expression.*

To engage in this form of coitus, *not nearly* the effort should be made to arouse the sexual passions of either of the parties, as has already been described as fitting for complete coitus. *The orgasm is not the desideratum in this case, but it is just a delightful expression of mutual love. It is a sort of prolonged and all-embracing kiss, in which the sex organs are included as well as the lips. They* kiss each other, as the *lips* kiss each other. It is "courting," par excellence, without the hampering of clothes or conventionality of any kind.

In this act, the lovers simply *drift,* petting each other, chatting with each other, visiting, loving, caressing in any one or all of a thousand ways. The hands "wander idly over the body," the husband's right hand being specially free and in perfect position to stroke his wife's back, her hips, her legs, and pet her from top to toe.

As this part of the act continues, it is the most natural thing in the world that the sex organs should tumesce, and that there should be a flow of both prostatic and pre-coital fluids. That is, the organs quietly and naturally make themselves ready for meeting. And when they are duly tumescent, are properly enlarged and lubricated, let the wife come over into her lover's arms, IN THE SECOND POSITION described, and the organs be slipped together easily, delightfully, and then, *let them stay so,* fully together, *but do not go on with the third part of the act,* the motion of the organs. Just lie still and enjoy the embrace, kiss, chat, court, love, dream, enjoy!

This union can be protracted to almost any length, after the lovers learn how to do it. Sometimes the organs may be together only a few minutes, sometimes for an hour, or even longer. If the parties get tired, or sleepy, part the organs, kiss good-night, and go to sleep. Although it is not at all uncommon for such lovers, who have fully learned this art, to go to sleep thus, in each other's arms, their sex organs united; and, in this position, have the organs detumesce, the penis grow limp and slip out of the vagina of its own accord, while the vagina also grows small and the clitoris subsides. This experience is most delightful and if once experienced, once well mastered by the husband and wife, it will continually grow in favor, to their mutual benefit.

This method is of special service during the "unfree time." If rightly used, it will not tend to increase the desire for "spending," but it will, on the contrary, allay and satisfy the sexual desires, most perfectly. If, while learning how, sometimes the inexperienced should "get run away with," and feel that it is better to go on and have the climax, all right. But, as time goes on, the practice of carrying the act only to the end of the *second* part, will grow, and in due time be well established. Those who have mastered this wholesome and loving art will sometimes meet in this way a score of times during a month or so, without once coming to the climax. Such meeting can be as often as the parties choose, and of as long, or as short duration as they elect. It is often an excellent way, to say "good-night;" and if, on waking in the morning, there is time before rising for a "little court," this slipping the organs together, for "just a minute," is a most excellent way to begin the day. The art is worth learning, and most people can learn it, if they try, *and are of the right spirit*!

To go back a little: In speaking of mutual masturbation on the part of the husband and wife, this method of satisfying the sex nature is of great value, sometimes, especially for use during the unfree time. If, during these two weeks, the parties get "waked up," and feel the need of sex exercise, they can satisfy each other with their hands in a way that will be a great relief to each. This is specially true for the husband; and a wife, who is enough of a woman to thus meet her husband's sex-needs, with her hand, when it is not expedient for him to meet her otherwise, is a wife to worship!

Sometimes, during the five days of menstruation, during which time the union of the organs is deemed not best, the wife can thus help her lover with her hand, to their delight and benefit. *Let love direct the way here, and all will be well.*

And here is a curious fact: The hand of the opposite sex will produce effects on the genitals of the other which will *not* be produced in any other way. Thus, a man may hold his penis in his own hand for a given length of time, longer or shorter, and no result will be effected, no secretion of prostate fluid be made, at all. But let his wife take his penis in *her* hand for the same length of time, and the flow of prostatic fluid will at once take place. This is true whether the penis be erect or detumescent. If the wife will hold her husband's limp penis in her hand for but a few minutes, even though the organ remains limp, the flow of prostatic fluid will take place! The same is true with regard to the husband's putting his hand on his wife's vulva. Should *she* hold her hand there, no pre-coital fluid would be secreted. With her husband's hand there, the flow would at once begin.

This is a remarkable physical and psychological phenomenon, and it is one especially worthy of note. It is this fact that makes *mutual* masturbation far superior to auto-erotism. A husband can thus satisfy a wife with his fingers, or a wife her husband with her hand, far better than either could bring himself or herself to the climax alone. This point is of great import, in considering many of the sex acts of husband and wife.

As a rule, let the husband and wife do *whatever*

their desire prompts or suggests, and just as they feel they would LIKE *to.* Only this, let all be in moderation. *Carry nothing to excess!*

Which suggests the question often asked: How frequently may coitus be engaged in? The answer is, just as often as is desired by *both parties, but never to the point of weariness or depletion of the physical, mental or spiritual body.* Use good sense here as elsewhere. We eat when we are hungry, but it is wrong to gorge oneself with food. The same rule holds with regard to sex exercise. *Satisfy the calls of nature, but* NEVER, *overdo the matter.* BE TEMPERATE, MANLY, WOMANLY! *Don't be afraid or ashamed to do what your desire and your best judgment say is right. Use common sense, and you will not go wrong.*

And don't wear each other out, either both together, or the one the other. Many men insist on their rights (THEY HAVE NO RIGHTS) and greatly debilitate themselves by excess of coition with their wives. Per contra, there are some women who wear the lives out of their husbands by the excessive calls they make upon them for sex-gratification. In the latter case, a man will "go to pieces" much faster than a woman who is over-taxed. To satisfy such a woman, a man must spend at least once every time his wife calls on him. This draws on his vital fluids, at every embrace; but, as has been stated, there is no escape of vital fluid from the woman, when she spends, and so she can reach and pass the orgasm, time and again, and still not have her vitality taxed. Indeed, in some cases, the oftener a woman spends, the more animated, robust and healthful she becomes. In case unmatched people meet as husband and wife, they should do their best to adjust themselves to each other's condition, keeping always in mind the best welfare, each of the other.

There are records of women who delight to spend a dozen times in a single night. One queen made a law that every man should cohabit with his wife at least seven times each night! Of course, she was an abnormal woman, though the author once knew a good orthodox deacon who would have been delighted to live under the rule of such a law, for seven times a night was the limit his wife imposed upon him! He was also abnormal.

Luther said twice a week was the rule for coitus, and this is a very common practice. No absolute rule can be given, however, except for each couple to act as they feel, keeping always within the bounds of common sense and true temperance.

There are some men and women so constituted, nervously, or by temperament, that they are *obliged* to rigorously *limit* their acts of coition. Some men cannot engage in the act more than once or twice a month and maintain their health. For them, the act draws on their vitality so severely that it quite upsets them, in almost every case. During the act, they are subjected to nervous shocks, they "see stars," and undergo rigors and nervous sweats which are severely debilitating. Often, too, they will lie awake all night after engaging in the act, and be more or less of a wreck for a day or two afterwards.

Some women, too, are of a similar nature of organization, and undergo similar experiences. Of course, in all such cases, unusual care should be taken never to reach the point of excess.

It is unfortunate if people are married who are ill-matched in this regard, especially so if the difference between the two is of a pronounced nature, as when the husband or the wife is very amorous and virile, while his or her mate is unable to engage in the act, to any considerable extent, without suffering therefrom. If such case arises, the best should be made of the situation, the more robust party accommodating himself or herself to the incompetency or inability of the other, and the weaker one doing all that can rightly be done to strengthen and develop his or her infirmity. If this is done, *the chances are many to one that, as times goes on, the parties will grow more and more alike – the strong becoming more docile and the weaker one more robust. Take time, love each other, court and be courted, and only the best results trill come of it all.*

Now there are some women who are called "anesthetic," that is, they have no sex-passion, though the sex parts may be normal. Many physicians declare that as high as forty per cent of the women *who are reared in modern social life* are thus lacking. These women engage in coitus, though they get no pleasure from the act. They never reach the orgasm, and have no sensation of delight from the act; they seldom secrete the pre-coital fluid, and hence the union of the organs, or their motion, are never easy or pleasurable. They can become mothers, and often such bear many children. Such condition is greatly to be regretted, and many women suffer greatly from this cause.

It is highly probable, though, that many women who are counted as thus lacking are *not, really, so!* Many women will begin married life wholly anesthetic, and, often, sometime will become normal in this regard. *This often happens. The probability is that many wives are not properly "courted" by their husbands* – THE FIRST PART OF THE ACT IS NEGLECTED, *or the husband merely acts on his rights* – cohabits like a goat, all in an instant, anxious only to gratify his own *lust*; and that, *under such treatment, the wife never gets a fair chance to really know her own powers.* Such cases are sad beyond telling. For the most part, *they are the result of ignorance on the part of the husband, and innocence and wrong teaching – wrong mental attitude – on the part of the wife.* HENCE THE NEED OF INSTRUCTIONS TO BOTH.

But if almost any woman will get the *right mental attitude* toward sex-meeting, and then can be courted, as has been prescribed in these pages, the cases are *rare indeed* where a woman can be found who is *really* anesthetic. If you, wife, or you, husband, are "up against" such a condition, try "courting," as herewith laid down, *in a proper mood and spirit, and you will come out all right. There is no doubt of it.*

On the contrary, if the man is "impotent" there is small hope of his ever coming out of such condition, and the chances are many to one that he will never be able to satisfy his wife sexually. He may be a "good man," in a way, but he can never be a good *husband,* in the full meaning of that word.

On the other hand, if a woman marries for money, or a home, or position, or place, or power, or a "meal-ticket" – for *anything but love,* she will doubtless be anesthetic *and stay so.* She deserves to! She sells herself for a mess of pottage, whoever she is. She may be a "good woman," but she can never be a good *wife.*

The question is sometimes asked as to how late in life the sex organs can function pleasurably and wholesomely for the parties concerned. And here, as elsewhere, the reply can only be that it all depends on the individual. But this is true, that, as a rule, the status of the individual during the years of active life will persist, even to old age, if the sex-functions are used and not abused. There is no function of the body, however, which will "go to pieces" quicker, and ever after be a wreck, as will the sex organs, if they are not treated rightly.

And this works both ways: If too rigorously held in check, *if denied all functioning whatever, the parts will atrophy, to the detriment of the whole nature, physical, mental, and spiritual.* The body will become "dried up," the sex organs shriveled, and a corresponding shrinking of the whole man or woman, in all parts of the being, is very apt to follow.

On the other hand, an excess of sex-functioning will soon deprive the individual of all such power whatsoever. A man will, in his comparatively early life, lose the power of erection, or tumescence entirely, as a result of excess, either by masturbation or from too frequent coitus; and on the part of the woman, many unfortunate conditions are liable to arise. However, for reasons that have already been stated, a woman who is strongly sexed, and of a pronounced amorous nature, can maintain even great excess of sex exercise without suffering such ill results as would befall a man who should so indulge. That is, an excessively passionate wife can far sooner wear the life out of a husband who is only moderately amorous, than can an abnormally passionate husband wear out a moderately amorous wife.

But if the sex nature of the husband and wife are well cared for during the years of active life, neither too much restrained or too profusely exercised, the functioning power of the sex organs will remain, even to old age, with all their pleasure-giving powers and sensations intact. This is a wonderful physiological fact, which leads to a conclusion, as follows: —

This fact of the staying qualities of the power of

sex functioning, even to old age, is the *supreme* proof of the fact that sex, in the human family, *serves a purpose other than reproduction*!

For, see! A woman loses the power to conceive when she reaches the "turn of life," when her menses cease, that is, when she is between forty and fifty years of age. And if pleasure in coition serves only to induce her to engage in the act for the purpose of increasing the probability of her becoming pregnant, if this is the *sole* purpose of desire for sex intercourse, such desire, such pleasure, *ought to cease* at that period of feminine life. *But this is by no means the case*! If a wife is a normal woman, sexually, and has neither abused her sex nature or had it abused, or neglected, and is a well woman, she will enjoy coitus as much after she has passed her three score and ten date in her life as she did before! She may not care to engage in the act as frequently as in her younger days; but if she is well courted by her old lover, all the joys of the former days are still hers, to as great a degree as ever. And what is true of her is true of her husband, if he is well preserved, as she is, has never abused himself or been abused.

This is a reward of virtue, for old lovers, that pays a big premium on righteous sex-action in earlier years! More than all, *it is a proof, beyond all question, that the purpose of sex in humanity is something more than procreation, that there is such a thing as the Art of Love, and that it ought to be taught and well learned by every husband and wife, in their early married life.*

X
CLEANLINESS

It would hardly seem necessary to be said, and yet many experiences of husbands and wives prove that it needs to be said, that both parties should take great pains to keep their bodies, all parts of them, always sweet and clean. Strange as it may seem, many wives are exceedingly careless in this respect! It is a matter of common report among men, that harlots take more pains to make and keep their bodies, and especially their genitals, clean and attractive, than many wives do! Surely, this ought not to be so, and yet it often is.

And that it is, is only one more unfortunate result that springs from the feeling of "Oh, we are married now." The wife or the husband feels that there is no longer any need of wooing each other. All of which leads to woe, woe, woe! The wife should keep her whole body so sweet and clean that her husband can kiss her from top to toe, if he wants to – and the chances are that he will want to, if she so keeps herself! In the one case, such a caress is a bit of heaven to a husband, in the other it is a bit of hell! It will disgust where it ought to delight. And when a wife disgusts her husband, the end of a happy married life has come!

The wife should always wash her vulva with soap and warm water before retiring, and if reservatus is to be engaged in in the morning, after urination, she should thoroughly cleanse the parts before union takes place. Let her be *ever* mindful to keep her "love cup" worthy to meet its lover.

And the husband should be equally careful to

keep his body sweet and clean. He should wash the glans penis thoroughly, with soap and water, at least once every day, drawing the foreskin back so as to fully cleanse the indenture above the gland, which secretes a substance that very soon emits an offensive odor unless removed. Both parties should keep their arm pits so that they will not be "smelly," and the feet should likewise be kept inodorous.

One of the chief objections to smoking or chewing tobacco is that it spoils the breath, and so makes it offensive to the wife, whereas it should be most attractive. In a word, both the husband and wife cannot be too careful, in all ways, in making and keeping their bodies mutually attractive. As has already been said, the sole aim of all the sexual experience of a husband and wife should be to raise the function more and more *away* from the plane of *physical* gratification and elevate it continually towards the realm of *mental* and *spiritual delight*. This is a mission of sex in the human family that should be made the most of. It involves the cultivation of the Art of Love, which is truly the art of arts, par excellence.

The secret of success in establishing righteous and happy sex relations between husband and wife is, on the part of the man, that *all his actions should be those of a loving gentleman*. This does not mean effeminacy on his part — he must be virile, bold, strong, aggressive, positive, *compelling*. And yet, all these manly virtues must be expressed in terms of *loving and gentle* ACTS. This is a paradox, but it is true!

On the part of the woman, the chief item on her side is, for her to attain a *correct mental and spiritual attitude toward her own sex-nature and that of her husband, and toward their common expression*. All her training and environment now hinder her from such achievement; but if she be a true woman, her nature will reveal the truth to her, and if she will trust to that — do what that prompts her to do, she will come out all right. It will take time to reach such results; but if she will persist, she will succeed. Let her come to the realization of the fact that sex in men and women is *not* unclean, vulgar, lowdown, sinful; but that it is *clean, pure, lofty,* GOD-BORN! Rightly exercised, it leads to the highest well-being of both the husband and wife; it brings them to their physical, mental and spiritual noblest and best. Let the wife get this view of the situation, which is the only true view, and then let her act accordingly, and she will have attained. A husband and wife who have reached this *modus vivendi* have established a heaven on earth.

EDITOR'S NOTE

Dr. Long's description of "Free Time" should be thoroughly understood by the readers of this book. Since it is practically impossible to conduct exact scientific tests under strict control (the reason for which can be readily understood) there is much difference of opinion among physicians and sexologists on this subject.

Some say there is no such thing as "Free Time." Others agree with Dr. Long that there is a period of "Free Time." Still a third group take the conservative viewpoint that further proof is necessary. The publishers offer this explanation as a necessary comment.

XI
Pregnancy

And now just a few words about having children, and this treatise will end.

As has already been said, every true husband and wife who are well enough and strong enough, and who are reasonably furnished with this world's goods, ought to have and rear at least two children. The world needs at least so many, even if all children lived and grew up, to keep up the constant number of people on the earth. But, far more than this, the husband and wife need children *to make a home complete, and a complete home is the supreme attainment of human life!*

This does not mean that people should not marry unless they can have children; there are many women who should never even try to become mothers. But these should not be deprived of all sexual joys for this reason. On the contrary, it is for their best good, in most cases, that they should marry and so live normal sex lives, in all respects except parenthood.

But, for the most part, husbands and wives *can* have children, if they so desire, *and they* SHOULD *so desire.*

And, so desiring, the question is, How can they best fulfil such desire?

As a matter of fact, there is very little that is really known about the begetting of children, and the securing of the best results from such action. The laws of human heredity are, as yet, for the most part, unknown. But common sense would seem to indicate a few things that must be best in the premises.

Thus, it would seem to be for the best that the husband and wife should be in good physical condition when a child is begotten. More than this, it would seem right that the act of begetting should be a *deliberate,* and not a mere *chance* begetting. Hence, in general, it is well for the husband and wife to *agree* upon a time for the begetting of a child, and *deliberately accomplish a sex-meeting for such purpose.* Although, one instinctively feels that such a deliberate meeting might be too matter of fact — too cold and formal, lacking in warm blood and genuine emotion; still, the probabilities are that even this could be overcome, if kept in mind and "provided for."

Referring to the things that have already been said, of course an embrace which is to result in pregnancy should be one of the most perfect that can possibly be experienced, one in which, in an ecstasy of love's delight, husband and wife merge their souls and bodies into a perfect oneness — it would seem that from such a meeting the best, and only the best results could come.

And so if the husband and wife will agree that from a given time on, they will cease to have a care to prevent conception; and then, sometime *immediately following the fifth day after the beginning of the menstrual flow,* they will naturally meet in a *perfect embrace,* the probabilities are that they will have done the best possible to secure the highest attainable results from the act of begetting a child.

As a rule, the proper time for such begetting is between the *fifth* and the *tenth* day after the beginning of the menstrual flow. It is sometimes best, however, to make the meeting earlier than this, even before the flow has ceased. Some women will conceive then who cannot do so at any other time. And so, if a wife should be unable to conceive between the fifth and the tenth day, as noted, let an earlier date be tried. If this should fail, consult a reliable physician.

It ought to be said, too, that putting off having children *too long*, is very apt to result in the sterility of the wife. Many a young wife, who has really wanted to have children *sometime*, and who would be greatly grieved if she thought she could *not* bear a child, has kept putting it off, and has done this *so often*, and for *so long*, that, when the "convenient day" does come, she finds that she has "sinned away her day of grace."

Speaking generally, the first baby should be born not much later than two years after marriage. There are, of course, exceptions to this, but it is a good rule to go by.

Have your children when you are young! This is common sense, it comes out best in the long run, and is the best thing to do, ninety-nine times in a hundred. Then, you are nearer the age of your children as they grow up than if you waited till you were in the late thirties before the children came. If your son or daughter is only twenty-some years younger than you are, you can be "kids" with them. If you are forty years old when they are born, you will always be "old folks" to them. Have the babies when you are young. It is far better so.

If no children come from the meeting of husband and wife consult a good doctor. But, in such event, if neither of the parties is to blame — or even otherwise, make the best of the situation, love each other, and make the most of wedded life with what is left.

Above all, with children or without (and a thousand times better with) make a home that is a home. That is what sex in the human family, what married life is for — to make a home. Nearly all that makes a home is centered around sex. No two normal *men* can make a home! No two normal *women* can make a home! *It takes a man and a woman to make a home. It takes father, mother and children to make the most perfect home. Make up your minds to have a most perfect home, and do your utmost to reach that goal*!

The query often arises in the minds of conscientious husbands and wives whether or not it is right to engage in coitus during pregnancy. On this point authorities differ, though most of them hold against such practice. The reasons they give for such adverse decision are all based on the same old infernal lie, namely, that, sexually, man is a mere animal, and so is subject to the laws and practices of mere animality. This is the worst outrage ever perfected by a false philosophy, which is heralded as the will of God. Out on it, altogether!

The simple truth, is that, if the husband and wife have *mastered the Art of Love*, so that they *mutually desire each other, and both long for sex exercise during the gestation period*, it is *perfectly right* and WISE for them to satisfy their *natural* COMMON wishes.

Of course, in such exercise, the utmost care should be taken not to press too hard upon the pelvic region of the woman, and in this regard, the word of caution needs to be heeded, as much by the prospective mother as by her mate. For, in the intensity of an orgasm, she may be tempted to crowd her body too violently against her husband, and so possible harm might result. Especially if the husband-superior position is taken during the act, he should be doubly careful not to permit the weight of his body to rest upon the enlarged part of the wife's anatomy, not in the least.

Indeed, the safest position for coitus, during pregnancy is, the woman on her back, and the man with his hips on the bed below hers, so that there is no possibility of pressure on her abdomen, which is perfectly free, in this position. In this position, the act may be engaged in, during pregnancy, as often as mutually desired, to the benefit of both parties.

Many pregnant women are more than usually passionate during the period of gestation. This is especially the case when the wife is happy in her condition, when she rejoices with exceeding great joy that she is on the way to experience the divine crown of wifehood — maternity! When such a woman desires her husband in love's embrace, it is cruel to deprive her of her longed-for delight.

Again, a wife, unpregnant, and when she rightfully wishes to remain so, may be somewhat fearful of becoming pregnant when she meets her husband, and so hesitate to give her passion full play, thereby missing the utmost delights of an embrace — but if she be pregnant, and so has no fear on this score, she can give herself up to utter abandonment to her impulses.

On this point, the final word is, use *common sense,* in a *spirit of absolute* MUTUALITY.

It goes without saying that it would be wicked, not to say a crime, for a husband to *compel* his wife to engage in coitus during pregnancy, against her will. On the other hand, many a wife has first experienced an orgasm when meeting her husband during pregnancy. The reason for this is that her fear of becoming pregnant is not then present — a condition which has before kept her from the climax.

It is further true that many a wife will greatly relieve and delight her husband if, on occasion, and as both may desire, she will relieve him with her hand; or sometimes, that they engage in mutual relief by this means during pregnancy.

XII
Conclusion

In closing this volume, the author wishes to say, as in opening, that no apology is offered for what has been written or said herewith. All has been set down in love, by a lover, for the sake of lovers yet to be, *in the hope of helping them on towards a divine consummation*.

As a final direction *Master the Art of Love,* which is *the divinest art in all the world; then study, and do your best to master the Science of Procreation*. It is these two, the Art of Love and the Science of Procreation, that, together, make married life a success. Without these, or, surely, without the first, there can be no such thing as true marriage. Hence, this is the *first* to learn, to master. It is worthy of the most careful study, the most faithful experiment.

It is right for people who never can have children to marry, and to share with each other mutual sex delights. It is far better for a husband and wife, having learned the Art of Love, to have children — and a home.

Thrice happy are the married lovers who live in the spirit of this sentiment, exalted to the highest spiritual plane; and if, out of such love exchanges children are begotten and born, and a perfect home is established, then married life is worth living. God has joined such together and nothing can put them asunder.

This volume is not something to be read once, and then put aside and forgotten. It should be studied, experimented upon, read again and again, especially by those who have difficulties in married life to overcome. And for *all* young married people, it should be a sort of Guide to Happiness that should be frequently consulted and its directions "tried out" and followed to the limit.

The fact is that, in true marriage, neither the husband nor the wife can be selfishly supreme. If selfishness asserts itself, on the part of either husband or wife, hell is sure to follow. There can be no true marriage under such circumstances, because there is no supremacy in true love, and it is only true love that can make an abiding true marriage. In true marriage, such as both God and Nature design should be, there is perfect comradery, equals walking with equals, with the principle of love and mutual helpfulness shared alike by both. Let no reader of this book forget these primal facts, or fail to act in accordance with them! For of such is the Kingdom of Heaven!